Y0-BCC-651

PHYSICS AND BIOLOGY
From Molecules to Life

WITHDRAWN
60984 81800

PHYSICS AND BIOLOGY
From Molecules to Life

Edited by

Jean François Allemand
Ecole Normale Supérieure, France

Pierre Desbiolles
Université Pierre et Marie Curie, France

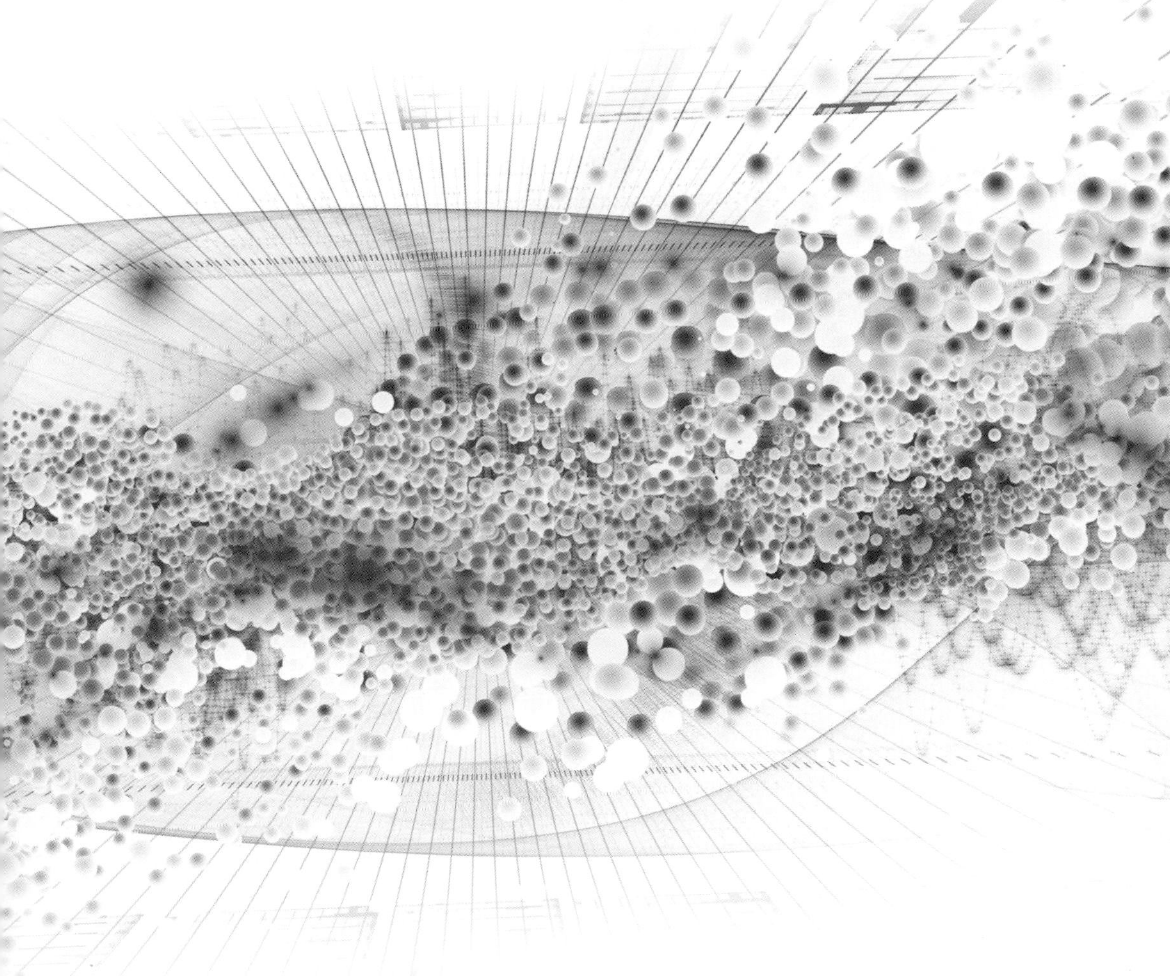

World Scientific

NEW JERSEY • LONDON • SINGAPORE • BEIJING • SHANGHAI • HONG KONG • TAIPEI • CHENNAI

Published by

World Scientific Publishing Co. Pte. Ltd.
5 Toh Tuck Link, Singapore 596224
USA office: 27 Warren Street, Suite 401-402, Hackensack, NJ 07601
UK office: 57 Shelton Street, Covent Garden, London WC2H 9HE

Library of Congress Cataloging-in-Publication Data
Allemand, Jean-François.
[Physique et biologie. English]
Physics and biology : from molecules to life / Jean-François Allemand (École Normale Supérieure, France), Pierre Desbiolles (Université Pierre et Marie Curie, France).
pages cm
Translation of French title: Physique et biologie : de la molécule au vivant, 2012.
Includes bibliographical references.
ISBN 978-9814618922 (hardcover : alk. paper)
1. Biophysics. I. Desbiolles, Pierre. II. Title.
QH505.A4713 2014
571.4--dc23
2014013887

British Library Cataloguing-in-Publication Data
A catalogue record for this book is available from the British Library.

Copyedited by Christopher Teo Qiang Long

Typeset by Stallion Press
Email: enquiries@stallionpress.com

Printed in Singapore by Mainland Press Pte Ltd.

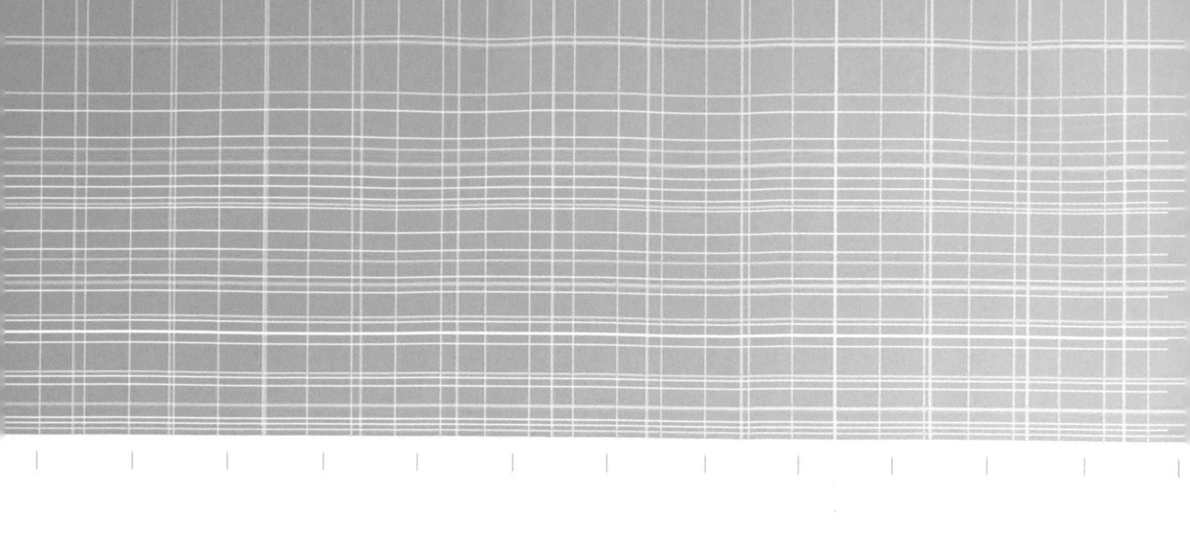

Coordinators and contributors

This book is a collective work. A dozen people did contribute to its writing.

Coordinators

Jean-François Allemand is professor at the École normale supérieure (ENS). He is a former student of the École normale supérieure de Cachan, with a doctorate from Pierre and Marie Curie university. He taught biophysics at the bachelors and masters level but also at the graduate level for physicists, chemists and biologists. His research activity takes place in the Statistical Physics laboratory of the ENS which focuses on the topics: DNA elastic properties, DNA molecular motors both *in vitro* and *in vivo* with micromanipulation and fluorescence techniques.

Pierre Desbiolles is a general inspector of the French ministry of education. He was professor at Pierre et Marie Curie University in Paris. His research activities were led in the Kastler laboratory at the École normale supérieure. He got a PhD in cold atoms physics, and in particular on Bose–Einstein condensation; he directed experiments, using fluorescence microscopy, to study DNA/proteins interactions at the single molecule level.

Contributors

— Olivier Benichou is a CNRS researcher working at Laboratoire de Physique Théorique de la Matière Condensée in Université Pierre and Marie Curie. His research is theoretical and focuses on statistical physics and random processes, and their applications to living systems.

— David Bensimon graduated with L. Kadanoff in the University of Chicago researching Chaos and non-linear phenomena. In 1986, he joined Bell Laboratories for a post-doctoral period studying convection in binary fluids. There, he met Vincent Croquette and together they established a research group at the École normale supérieure in Paris that moved towards more biologically inspired problems. Initially he studied the shape of membranes and vesicles and then together with V. Croquette and their students the mechanical properties of single DNA molecules, DNA/protein interactions at the single molecule level. Recently, D. Bensimon has diversified his interest developing means to control optically the activity of proteins in single cell of a zebrafish embryo. He is using that technique to study problems in development and cancer. He is also interested in the evolution and ecology of bacterial colonies.

— Yves Boubenec received the Biology master's degree from the École normale supérieure (ENS), Paris, France, in 2008 and the doctorate degree in Neurosciences from the ENS and UPMC, Paris, France, in 2013. From 2013, he joined the UNIC in Gif-sur-Yvette, France with research interests including whisker mechanics and tactile texture perception in rats. His focus is now on sound textures in the ferret auditory system.

— Laurent Bourdieu is a 44 year old CNRS researcher. After a PhD in soft condensed matter at the Institut Curie (Paris) and a postdoc in biophysics at The Rockefeller University (New York), he worked at the University of Strasbourg and now at the École normale supérieure in Paris, in the IBENS laboratory.

— Jean-François Léger is a 40-year old CNRS researcher. After a PhD in biophysics at the University of Strasbourg, he did a postdoc in neurophysiology at the University of Freiburg. He is currently working at the École normale supérieure in Paris, in the IBENS laboratory.

The Laurent Bourdieu and Jean-François Léger researches focus on study of the cellular and network mechanisms underlying sensory integration in rodent cortex using advanced two-photon microscopy.

— Vincent Croquette followed the ESPCI engineer school in Paris; he passed his PhD in 1986 on Non-linear physics studying hydrodynamics and transition to chaos in dynamical systems. After entering CNRS and a Postdoc in Bell Labs, he joined ENS with D. Bensimon to pursue Non-linear dynamics. In 1992, both of them changed their research interests to biophysics. Since then he has pioneered single molecule micromanipulation to investigate molecular motors using magnetic tweezers.

— As a graduate student, Maxime Dahan was trained in quantum physics in Pr. Cohen-Tannoudji's lab at École normale supérieure. After a post-doctoral visit at Berkeley under the supervision on Dr. Shimon Weiss, he was recruited as a CNRS scientist. From 2000 to 2012, he worked as a group leader at École normale on the development and application of single molecule assays in cell biology, using a combination of optical tools and new nanotechnologies. Since 2013, he has been head of the Laboratoire Physico-Chimie at the Institut Curie.

— Georges Debrégeas is a Director of Research at the CNRS. He currently works in the Laboratoire Jean Perrin, within the University Pierre and Marie Curie in Paris. His background is in statistical physics and mechanics. In the last 5 years, he has worked on the biomechanical and neuronal substrates underlying mechanosensation.

— Pascal Martin is currently Director of Research at CNRS. He did his undergraduate studies in Physics and Chemistry at the "École Supérieure de Physique et Chimie Industrielles de la ville de Paris (ESPCI)" and obtained in 1997 a PhD in Physics from the University Pierre and Marie Curie (Paris). He leads a research team at the Laboratoire Physico-Chimie Curie in the Curie Institute (Paris) that works on the active mechanosensitivity of hair cells from the inner ear and of purified molecular-motor assemblies in *in vitro* motility assays.

— Lia Giuseppe was a researcher at the Centre de Génétique Moléculaire (CGM institute) at the CNRS in Gif-sur-Yvette. He is a former student of the Universita degli Studi di Milano in Italy, and doctorate from Denis Diderot-Paris VII university France.

— Sylvie Hénon is professor at University Paris Diderot. Her background is in soft matter physics. She has been working on cellular mechanics and mechanotransduction for about 15 years, mostly by micromanipulations coupled to fluorescence microscopy.

— Terence Strick studied physics as an undergraduate at Princeton and obtained his PhD in Biology while working in the group of David Bensimon and Vincent Croquette at the Statistical Physics Laboratory of the École normale supérieure in Paris. His main interest is the study of protein-DNA interactions involved in DNA transcription, replication, repair and chromosome structure.

— Cécile Sykes was trained as a solid state physicist (PhD in Semiconductor Physics), and got interested gradually in Soft Matter Physics and in Biophysics. She founded a group in 2001 at the Curie Institute in Paris named "Biomimetism of Cellular Movements" that aims at designing and studying biomimetic systems for a better understanding of cell motility and cell shape changes. The motivation of her work is to understand how cells move and change shape in a disease like cancer.

— Raphael Voituriez is a CNRS researcher working at "Laboratoire Jean Perrin and Laboratoire de Physique Théorique de la Matière Condensée" in Université Pierre and Marie Curie. His research is theoretical and focuses on statistical physics and random processes, and their applications to living systems.

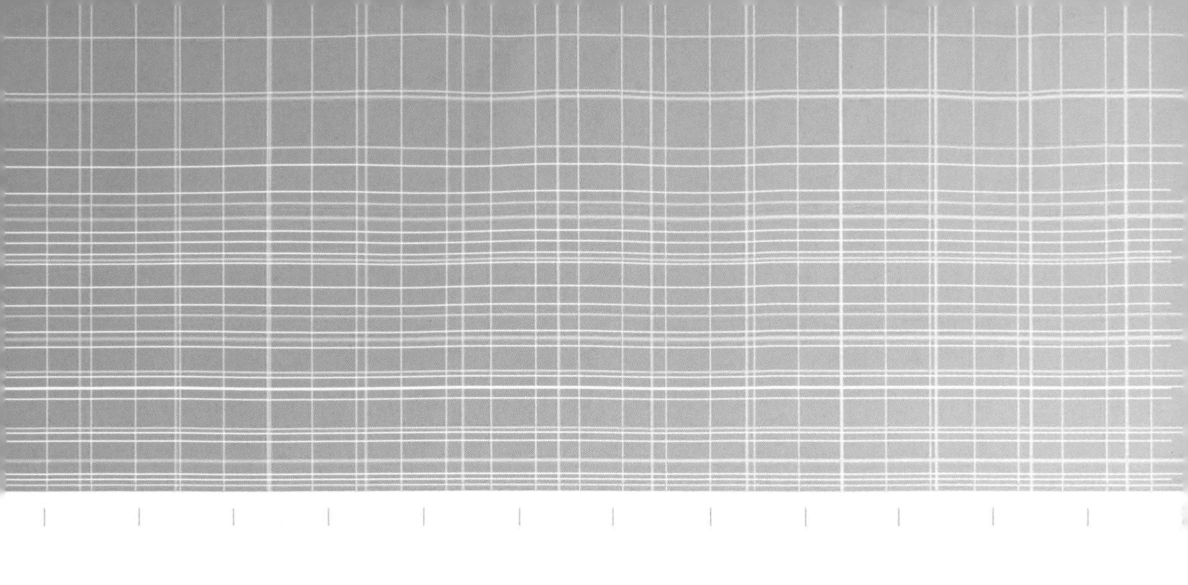

Table of contents

Preface

Jacques Prost, *CNRS Researcher, member of the French Academy of Sciences, ESPCI ParisTech director.*

Physicists have been interested in biology for a long time. In several chapters of this book the historical contributions of Hooke and Vans Leeuwenhoek to biology in the seventeenth century are rightly mentioned. The identification of the cell as the building block of the living world results directly from the development of the optical microscope they pioneered. More recently, during the twentieth century, physicists contributed to about one major experimental contribution per decade and one conceptual contribution every fifteen years. The explosion of the research activity at the interface between physics and biology over the last twenty years thus follows a long tradition of interest of physicists in the living world. But the scale of this investment is new. From the femto-second in the study of mechanisms of bio-catalysis to the geological time in the modeling of evolution processes, from the fraction of a nanometer in protein structures to macroscopic scales in developmental biology or to the continental scales of population dynamics, physicists have shown an insatiable curiosity. They also proved that a physicist's approach is useful when it goes with a good collaboration with biologists, as explained and practiced very well by the coordinators of this work. Michèle Leduc and Michel Le Bellac were right to ask Jean-Francois Allemand and Pierre Desbiolles to coordinate a book accessible to everyone and that can illustrate the impressive results that have been accumulated these past twenty years at the physics-biology interface. The explosion of the activity and associated success deserve to be popularized.

I can imagine the hesitations of Jean-François and Pierre upon choosing the content of the book. Some omissions were obvious: who now realizes that the electrophoresis, centrifugation, mass spectrometry were once research subjects in physics? These techniques have become daily tools in biology. The same goes for electron microscopy, crystallography and nuclear magnetic resonance: they became part of "structural biology". The beautiful ultrafast optical experiments designed to study the mechanisms of intimate biocatalysis could have been chosen, but eminently physical, they seek to answer more "chemical" than physical questions. Jean-François and Pierre could also have thought of the adaptation processes in biology, the physics of protein networks, statistical physics of folding proteins or RNA, evolution, bacterial resistance to antibiotics, the physics of tissues, etc. The list is long.

Yet the choice of Jean-François and Pierre is logical and consistent. Two types of studies have seen particularly dramatic evolutions during the last two decades: the single molecule mechanical studies and physical approach of the cell. It was then natural to introduce the concepts needed to understand these aspects: basic concepts of cell biology, leading technologies starting from fluorescence microscopy, which has contributed so much to biology. Since the discovery of "green fluorescent proteins" biologists know how to use them as a "reporter" for studying specific mechanisms. The presentation, while remaining very affordable, describes the latest developments that give access to information well below the theoretical limits of the optical resolution of microscopes. But the reader must be reassured, the laws of physics are met! With this new knowledge, it is natural to try to understand the single molecule experiments. In fact the first single molecule experiments have been carried out on membrane ion channels in the 1970s. Significant improvement of the measurement of low currents opened this field of investigation. Similarly, in the 1990s, measurements of weak forces, typically in the pico-newton range, and visualization of single molecules have paved the way for a series of great experiments on the exceptional mechanical properties of molecules. Again, the presentation, while keeping the simplicity required for a wide audience, goes to the most recent technical developments in the field. The example of DNA mechanical properties shows the high degree of knowledge that one is now able to obtain on this molecule, which is the custodian of our identity. And this is not an esoteric game for physicists! We need to know all these properties if we want to understand the exquisite mechanisms of life! The next logical step is then to use single molecule experiments to study "molecular motors". These molecules are real engines: they consume chemical energy, and are able to provide mechanical work. Some, like the spectacular rotatory motor named F1 ATPase can also work in reverse, and it is even its natural physiological function: it converts mechanical energy into chemical energy. The RNA polymerase provides both mechanical work and causes very specific chemical changes. The concepts of thermodynamics apply quite well, provided that we use the correct averaging, and the study of fluctuations proves to be very rich. If one of the engines described in this chapter is deficient, that's life which disappears! This field of engines and of "machinery" active on DNA and RNA has taken a great importance and several chapters could have been devoted to this subject. Jean-François and Pierre have kept a reasonable size while providing a good measure of the importance of the addressed issues. It was then natural to go to the cellular level. If the engines are extremely complex, the cell is infinitely more. In continuation of the previous chapters, the authors have chosen to focus the scope of the book on the mechanical properties and cellular movement, as both are more accessible to the physicist. They introduce the intricacies of a world where viscosity dominates (life at low "Reynolds number") and those of the cytoskeleton with or without engines! The cell as a "homunculus"? Not quite, but it contains all the information necessary for the formation of an individual! From the cell to the organ: is there something nobler than the brain? The choice was again quite natural. I remember Pierre-Gilles de Gennes, explaining in the preamble of a remarkable lecture, on the functioning of the brain, that many famous physicists investigated in vain this subject in their old years. Here, physicists in the prime of life describe the spectacular progresses achieved in recent years in the knowledge of the brain! After reading this chapter you will be convinced that soon we will understand how it works. Most of the brain processes information conveyed by peripheral detection systems. The most sensitive of them are the auditory system and the surface shape detection system, skin for primates, whiskers for rodents. You will learn that the ear is not only a sound detector but also an emitter, that it is strongly non-linear and that it can detect better than its own noise! You will also learn why the fingerprints are essential for assessing a surface roughness and what whiskers have in common with fishing rods! Mainly oriented towards experiments, it was necessary that this book had at least one chapter describing a theoretical advance. Again there were many possible choices. Optimization of search strategies from molecular to animal scales is a good example of the use of statistical physics to concrete problems and it illustrates very well the type of concepts that theory introduces. After reading this chapter you will not go search your lost keys under the only shining street lamp! You will take example from your dog, running in one direction with conviction, and seemingly random rummaging for some time and so on! I'm not sure this technique replaces a good memory, but you will see it is very effective on many scales. I encourage you to dive into this book. You will find a strong body of information essential to the understanding of life and I am convinced that you will be even more eager to learn after reading it. It was designed exactly for this purpose, I think.

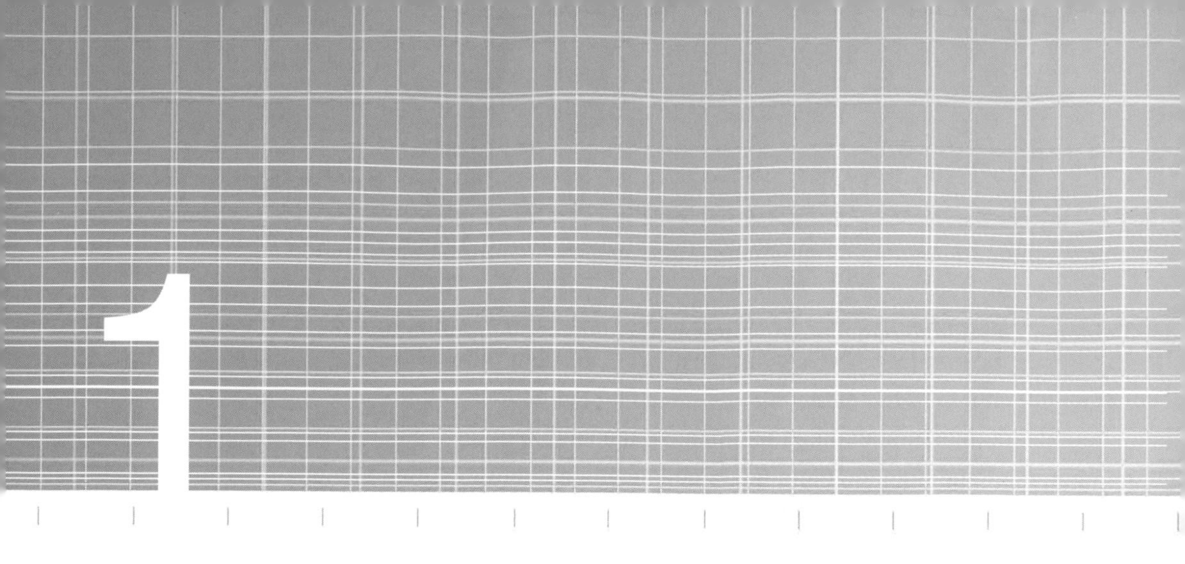

Some biology basic principles

Giuseppe Lia, *PhD in Biology*
Terence Strick, *CNRS Researcher, Institut Jacques Monod, Paris.*

1 Introduction

The goal of this first chapter is to facilitate the discovery of what could be for a new reader a new world: biology. Without pretending to be exhaustive, this first chapter introduces basic concepts useful to situate the subjects treated in the next chapters in the landscape of current knowledge. This first chapter starts with the description of the cell, the basic unit of all organisms, before moving on to the molecular level and introducing DNA, the medium in which genetic information is written.

Even if the complete reading of this chapter is not essential for understanding the following chapters, which are all independent and where most technical terms are defined, it will allow readers to contextualize its contents and to place them in a biological perspective.

By consequence, the first part, which describes the cell, is useful to approach Chapters 2, 5 and 7, and the second part, which concerns DNA, is useful to approach Chapters 3, 4, 7 and 8.

The reader less familiar with biology will need time to assimilate not only the concepts but also the vocabulary of biology and biochemistry. This reader should not hesitate to stop reading, take a pause, and return to difficult concepts at a later point.

Interdisciplinary work requires an effort in assimilation of new vocabulary and different modes of reasoning: it is one of the main difficulties but also the richness of this kind of research.

2 At the cellular level

2.1 General cell structure

The cell is the basic structural and functional unit of all living beings. The word cell comes from the Latin word *cella*, which means empty space. It was used for the first time in 1665 by Robert Hooke, the first scientist to observe the plant cell using a rudimentary microscope. However, we have to wait for the middle of the 19th century to assert that all living being are constituted by cells.

Each cell is a defined entity. It is isolated from the other cells by a cell membrane, which sometimes is rigid like in plants or in bacteria, in which we talk about a cell wall. Inside the cells there are a lot of chemicals and subcellular structures which make possible cell functions.

The region where genetic material is located is delimited by a membrane in organisms called Eukaryotes (like animals). This region is called the nucleus and its membrane, the nuclear envelope. In simpler organisms, such as Prokaryotes (e.g. bacteria), there is no nuclear envelope and the region where the genetic material is located is called nucleoid. The enviroment outside the nucleus or the nucleoid is called the cytoplasm.

The cytoplasm contains all the machineries necessary for cell function (Fig. 1.1).

The cytoplasm contains all of the chemicals located inside the internal part of the cell membrane. This is the main functional region of the cell, the location where most of the cell functions take place. In Eukaryotes it is constituted by three main elements: cytosol, organelles and inclusions, which are now discussed.

— Cytosol is a viscous liquid in which the other elements constituting cytoplasm are suspended. It is mainly constituted of water, and it contains soluble proteins, salts, and several other solutes.
— The inclusions are not functional elements but chemicals which can be present or not, depending on the cell type (like for example glycogen, pigments, metals, etc.)
— The organelles are the cell's metabolic apparatus. Depending on the presence or absence of internal organelles, like the nucleus, we classify the living organism

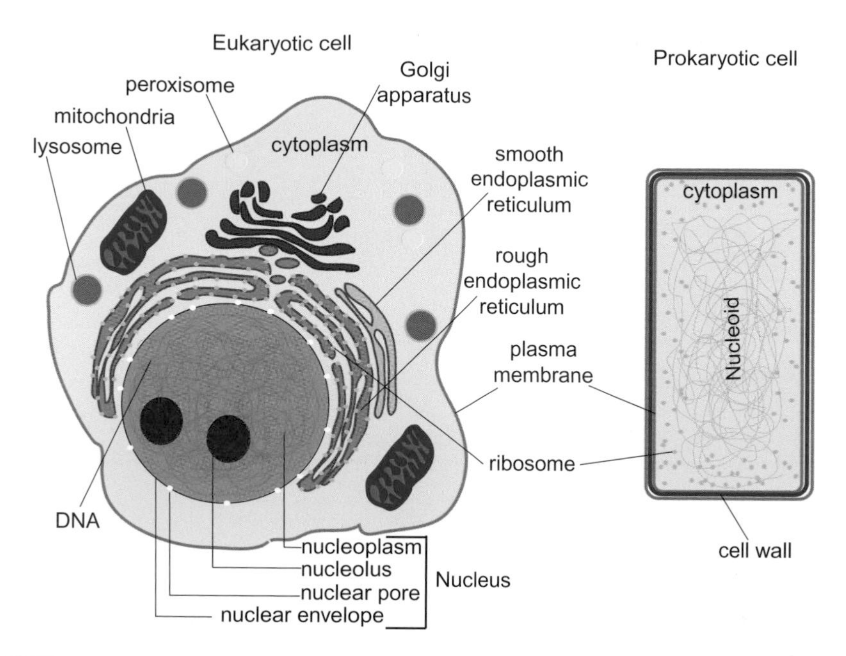

Figure 1.1. Schematic view of the typical organization of a eukaryotic cell and its multiple compartments, in particular the nucleus which contains the genetic material, and of a prokaryotic cell which has no nucleus.

in two categories: the Prokaryotes (Eubacteria and Archaea) and the Eukaryotes (Plants, Fungi, Animals and the rest of Eukaryotes called Protists).

Prokaryotic cells lack internal compartments, and by consequence the genetic material is in direct contact with cytoplasm. The cell membrane which delimits the cell is reinforced with a layer of peptidoglycans (molecules which are constituted by a peptidic and a glucidic region). This layer constitutes the cell wall and allows bacteria to resist osmotic shock, i.e. pressure variations due to concentration differences between the inside and the outside of the cell.

In Eukaryotes, the compartments inside the cell are delimited by a membrane, the composition of which is similar to that of the cell membrane, but sufficiently different nonetheless that the delimited enviroment can be differentiated from the surrounding cytoplasm. This partitioning is crucial for numerous cell functions: without it, all the enzymes will be randomly mixed and the biochemical activities will be totally disorganized.

2.2 Cell membrane

The cell membrane is a really thin structure (7–8 nm thick), constituted by a double layer of lipid molecules which includes some dispersed proteins (Fig. 1.2).

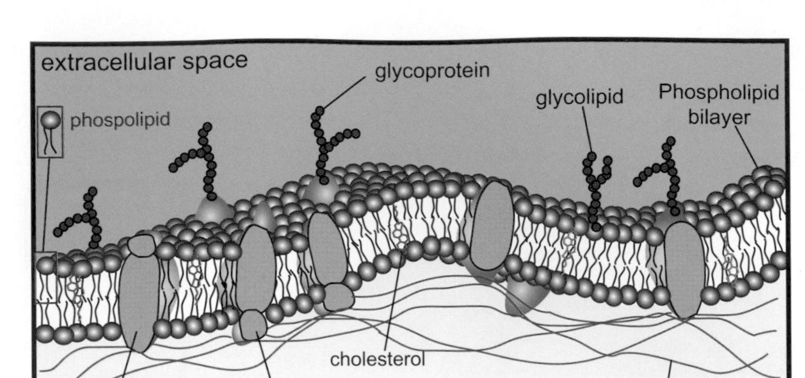

Figure 1.2. Schematic view of cell membrane components inside the two lipid bilayers.

In general the cell membrane is constituted of lipids (40%) and proteins (60%). The lipid bilayer is mainly composed of phospholipids (lipids bearing a phosphate ion) and it is relatively impermeable to most water-soluble molecules. Phospholipids are amphiphilic molecules: they have a polar extremity, called heads and which contains a phophate group, and a non-polar extremity, constituted by two long fatty acid hydrocarbon chains, called tails.

The polar head is hydrophilic and is attracted to water, while the non-polar tails are hydrophobic and are repelled by water and charged molecules. This specific property of phospholipids is responsible for the organization of all biological membranes, called "sandwiches". In this organization, molecules are disposed in two layers with the two hydrophobic tails facing each other inside the membrane, while their polar heads are exposed to water inside and outside the cell. This spontaneous organization of phospholipids allows biological membranes to automatically assemble into closed structures, mainly spherical. Almost 10% of phospholipids facing outside the cell are linked to glucids (sugar molecules); these are called glycolipids. Membranes also contain large amounts of cholesterol, which is disposed between phospholipid tails thus stabilizing membranes.

The cell membrane is a very dynamic, fluid structure. Lipid molecules can diffuse laterally, but polar–non-polar interactions prevent them from turning around ("flipping") or passing to another fat lipid layer. There are two distinct populations of membrane proteins: transmembrane proteins which are inserted through the lipid bilayers, and peripheral proteins, which are located on the intracellular or extracellular side of the membrane. Some proteins freely diffuse within the membrane, while others, like peripheral proteins, are more limited in their movement and appear as if "anchored" to some internal cell structures which constitute the cell cytoskeleton.

Each cell has to find in its surrounding environment the precise amount of chemicals necessary to live, and at the same time it has to prevent the entry of all chemicals in excess. For this reason the membrane forms a selective, permeable barrier. Movement across the membrane of chemicals can be active or passive. In passive movement, molecules pass across the membrane without the input of cellular energy and they follow their concentration gradient or electrical potential (in the case of ions). Non-polar and liposoluble molecules can diffuse directly across the lipid bilayer (chemicals like oxygen, carbonic gas, fat and alcohols). Polar and charged molecules can diffuse across the membrane if they are small enough to pass through the holes formed by channel proteins. Water and molecules with a diameter smaller than 0.8 nm can pass across the membrane by this mechanism.

In the case of larger molecules, there are some channel proteins which guarantee the entry of ions and specific molecules (we speak in those cases of facilited diffusion). Since a lot of molecules cannot pass through a cell membrane, for example intracellular proteins and some ions, there is a pressure difference between the two sides of the cell membrane called the osmotic pressure. Generally, osmotic phenomena, which cause important modification of the cell tonus, proceed until equalization of all pressures acting on membrane (osmotic and hydrostatic pressure). In active transport mechanisms, the cell uses metabolic energy (ATP) to transport some molecules despite their concentration gradient and electrical potential through a cell membrane, using as transporters some protein "pumps" (Fig. 1.3). The most studied active transports are the sodium pump, the potassium pump, the calcium pump, and the proton pump (see Chapter 4).

This differential permeability and the active transport of ions generate an electrical potential difference between the two sides of a membrane, called the membrane potential. In a state of rest, every cell of the organism mantain a non-zero membrane potential, which is between −20 and −200 millivolts (mV) depending on the organism and cell type.

For this reason, all cells are said to be polarized. This voltage (or charge separation) exists only at the membrane level. If we could sum all the positive and negative charges inside the cytoplasm, we would notice that the inside of the cell is electrically neutral.

2.3 Internal organelles

Eukaryotic cells are constituted of different organelles, which we will now describe in detail (Fig. 1.1).

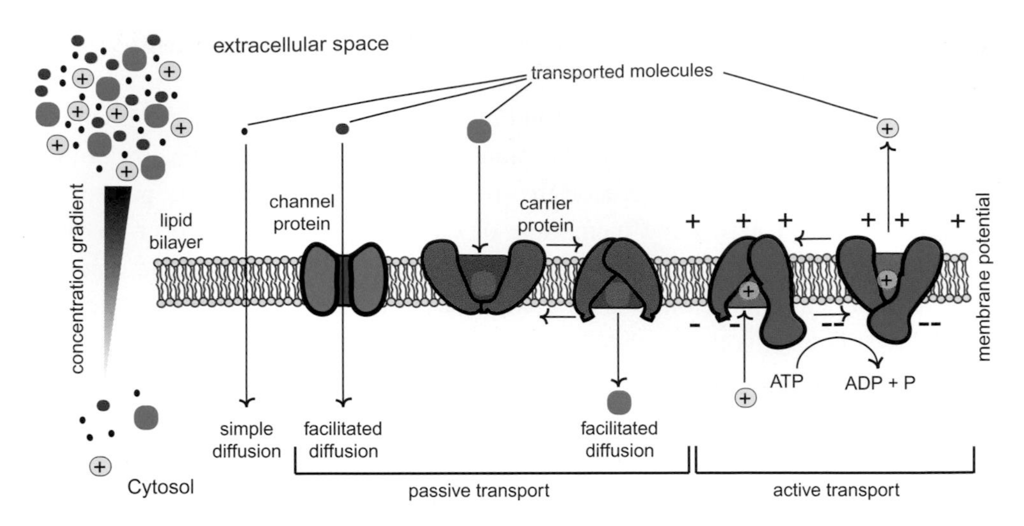

Figure 1.3. Schematic description of various types of transporters across the cell membrane. Passive transport is carried out always according to a concentration gradient or an electrical potential, simply by diffusion through the lipid bilayer or by facilitated diffusion mediated by channel proteins or carrier. Active transport uses the energy of ATP hydrolysis to transport molecules against their concentration gradient or electrical potential. (Inspired by Figure 11-4 in Molecular Biology of Cell, Garland Publishing Inc.)

The nucleus

The nucleus, which has a diameter of 5 μm, is the biggest organelle of the cell. It can be considered as the "safe" of the cell, because there lies inside the genetic material, materialized by the DNA molecule. It is composed of three regions or distinct structures: the nuclear envelope (or nuclear membrane), the nucleolus and the DNA molecules. The nuclear envelope is constituted by the superposition of two lipid bilayers. The external side of the nuclear envelope extends the endoplasmic reticulum (see later) and it is covered with ribosomes on the external side. In some points, the two membranes of the nuclear envelope merge, forming a nuclear pore. Nuclear pores allow exchange between the nucleus and the cytoplasm. Inside the nuclear envelope we have the nucleoplasm in which the nucleolus and DNA are suspended. Placed end-to-end the DNA molecules inside the nucleus are very long compared to the nucleus size (2 m long for all the DNA of a human cell). Consequently, DNA has to be compacted under the form of chromatin. During cell division, chromatin is condensed and structured, forming chromosomes. Each chromosome is a unit which contains part of the genetic information (in humans, the entire genetic content of the cell is stored in 23 pairs of chromosomes). Inside the nucleus we can also distinguish spherical particles called the nucleoli, which are the assembly regions of ribosomal subunits. As we will see later, ribosomes are central to the cell as they are the subunit which synthetizes proteins from RNA. The nuclear membrane is linked to the membrane of the other organelles, forming an interactive intracellular network called the endomembrane system.

2.4 The endomembrane system

The endomembrane system is an intracellular network of membranes located between the nucleus and the cell membrane. It is constituted of endoplasmic reticulum, the Golgi apparatus, peroxysomes and lysosomes.

The endoplasmic reticulum (ER) is an extended network of interconnected pipes and parallel membranes which are intertwined and curled inside the cytoplasm forming enclosed regions called "Cisternae". The ER extends the nuclear envelope and represents almost half of the cell membrane. There are two types of ER: the rough ER and the smooth ER. The rough ER owes its name to the presence of ribosomes on the external side. The proteins assembled by ribosomes are inserted inside the cisternae composing the rough ER. The smooth ER extends the rough ER and it is constituted by an interconnected network of tubules. It doesn't present any cisternae. The smooth ER enzyme and the enzymes of the rough ER catalyze reactions for lipids metabolism, like cholesterol synthesis and lipid region of lipoproteins. Inside the muscle cells of skeleton and heart tissue we find a very complex smooth ER, called "sarcoplasmic reticulum", which plays a very important role in stocking and releasing calcium ions during muscle contraction.

The Golgi apparatus is an endomembrane network which is similar to the endoplasmic reticulum, a stack of flat membranes surrounded by small vesicles. This network of membranes directs most of the "protein traffic" inside the cell. Its principal function is to chemically modify proteins by adding certain chemical groups (sugar, phosphate, sulfate...), in what is known as post-translational modification of protein. Those modified proteins are later concentrated, packed, and organized inside a membrane depending on their final destination. The Golgi apparatus also produces vesicles containing transmembrane proteins and lipids which are addressed to cell membranes or other organelle membranes.

The Peroxysomes are membrane bags (vesicles) containing oxydases, enzymes which use molecular oxygen (O_2) to neutralize multiple toxic and dangerous substances, like alcohol and formaldheyde, and to oxidize some long-chain fatty acids. The most important function of peroxysomes is to defuse free radicals, reactive chemical substances with unpaired electrons which can wreak havoc in the structure of proteins, lipids, and nucleic acids. Peroxysomes attack free radicals like superoxyde ion (O_2^-) and the hydroxyl radical (–OH) transforming them into hydrogen peroxyde (H_2O_2). Then the enzyme called catalase reduces the hydrogen peroxyde into water.

The Lysosomes are spherical vesicles containing digestive enzymes. The enzymes contained therein can digest all kinds of biological molecules, and they work in an acidic enviroment (pH 5). To generate such an enviroment, the lysosome membrane contains hydrogen ion (proton) "pumps" which allow the accumulation of hydrogen ions coming from the cytoplasm. We can consider the lysosomes like the "demolition site" of the cell. In fact, they guarantee the digestion of particles ingested by endocytosis, like bacteria, toxins and virus. They also allow degradation of old organelles which are no longer functional.

The mitochondria

The mitochondria are organelles surrounded by two membranes, and the internal membrane has the particularity of being highly folded upon itself. Mitochondria, which are present in all eukaryotic cells, are organelles which produce the energy necessary for cell survival. Cell components use generally only one source of energy: ATP (adenosine triphosphate). The mitochondria's role is to transform the energy furnished by the cell under the form of organic food material (glucids, lipids, proteins) into ATP which can be dispensed and used throughout the cell. This is produced via a proton pump, the F1-ATPase which will be studied in more detail in Chapter 4.

2.5 The cytoskeleton

The cytoskeleton is a complex network of protein filaments and tubules extending throughout the entire cytoplasm. This network helps maintain cell shape, supports and anchors cellular structures and organelles, and produces the different types of cell movements. It is, at the same time, the cell "skeleton" and "muscle". The cytoskeleton is a dynamic system: it assembles and disassembles, it reorganizes itself continuously during different cellular events (migration, cell division, etc.). Each cytoskeleton fiber is built by polymerization of identical protein subunits which are assembled, in some instances consuming energy to do so, into elongated structures. Depending on the nature and the role of protein forming the fiber, we can distinguish three main kinds of protein structure which compose the cytoskeleton: actin filaments or microfilaments, intermediate filaments and microtubules.

Actin filaments

Actin is a globular protein which assembles forming filaments (Fig. 1.4). It is the most abundant protein in large animal cells (almost 5% of total protein mass in

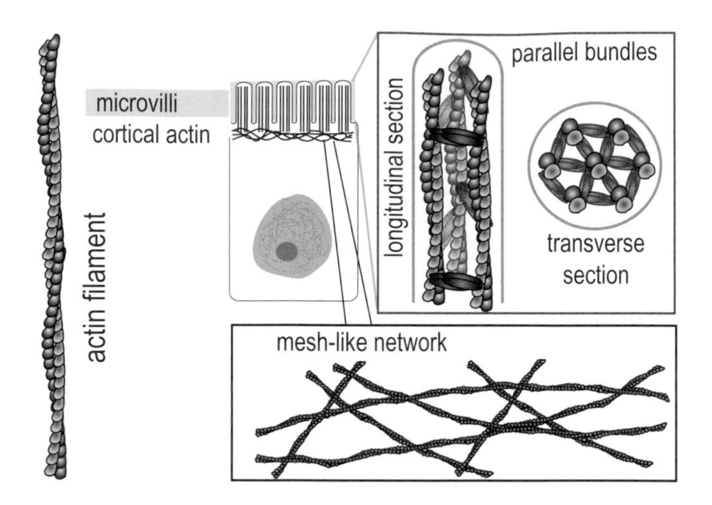

Figure 1.4. Schematic showing the two modes of organization of actin. Inside the microvilli (small protrusions of cell membrane which increase the absorbing surface), actin is organized in a parallel network. Under the cell membrane, actin fibers are disposed in network forming meshes.

muscle cell). Actin polymerization requires the energetic contribution of ATP. The actin polymer is a compact helix of 5–9 nm diameter forming a polar filament, which means the fiber has an orientation and the two extremities are not equivalents. The actin filament is also rather rigid — for instance thermal agitation can deform it from a straight line only if it is longer than a few microns. Actin filaments form dynamic structures which are stabilized by the proteins which can associate with them. Most microfilaments contribute to the motility (the ability to move) or changes in cell shape. Actin polymerization is sufficient for propelling bacteria as we will see later on in Chapter 5. Actin filaments can also organize into different structures depending on their position inside the cell. Inside the microvilli (Fig. 1.4), small protrusions of cell membrane of typically 1 μm are formed in which actin fibers are organized into parallel bundles and play a structural role — these can be found inside the intestine for instance. The peripheral actin network under the cell membrane acts as a support for polymerization of new filaments which push the cell membrane outward, forming lamellipodia, cellular bulges used by the cell to explore the surrounding environment (Fig. 1.5). Under the cell membrane and inside lamellipodia, actin filaments are disposed in a network forming meshes called cortical actin, the role of which is to support and reinforce the cell surface. Inside contractile fibers such as muscle fibers, the contraction mechanism is based on the sliding of actin filaments nested with type II myosin, a motor protein which uses the energy of hydrolysis of ATP to cause the sliding of actin fibers one relative to the other (see Chapter 4 and 5). This interaction between actin and myosin produces the contraction force of

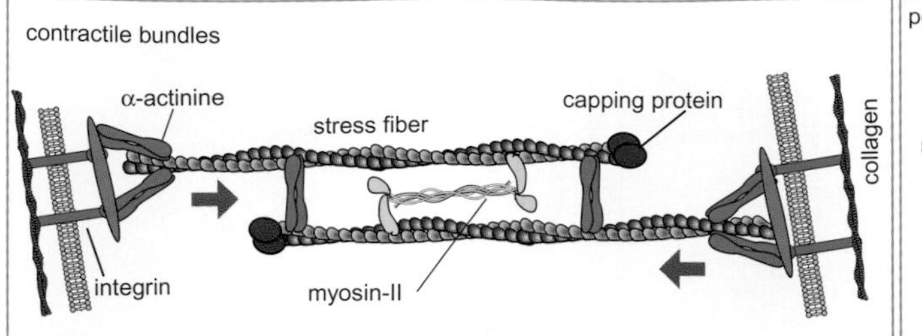

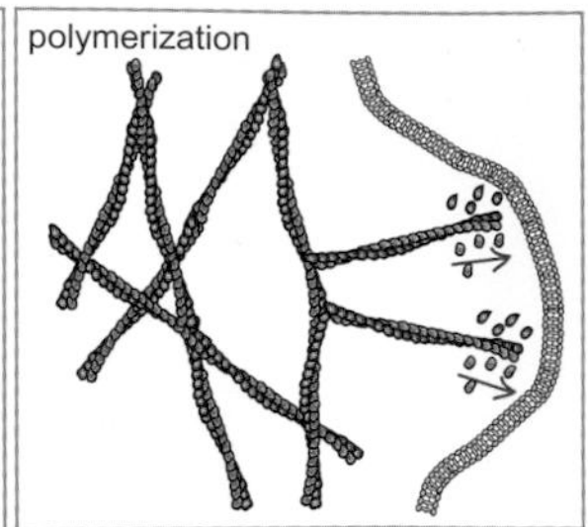

Figure 1.5. Generation of forces mediated by actin fiber. Inside the contractile fiber (disposed in a parallel structure via the alpha-actinin) myosin-driven sliding of two adjacent tensile fibers generates force on the plasma membrane, where fibers are anchored at the level of integrins (Fig. 1.6). At lamellipodia, polymerization of actine fibers likewise pushes the membrane.

muscle cell, and also governs the formation of the contractile ring which separates the cell during cell division.

Intermediate filaments

Intermediate filaments are protein polymers with a secondary structure in the form of a 10 nm helix. They are present inside the cytoplasm in most cells (Fig. 1.6). They are called intermediate because their apparent diameter is located between the diameter of actin and that of microtubules. The filaments are organized following the structure of a twisted rope. They possess significant resistance to tension, and they constitute the most stable and permanent elements of the cytoskeleton. Once formed, intermediate filaments are stable and they do not dissociate. Unlike the other two types of filaments, they are not implicated in cell movement, but act as an internal shroud which resists stretching forces acting on the cell. They also contribute to formation of desmosomes and hemi-desmosomes, structures connecting cells to each other and to the basal lamina, an assembly of extracellular proteins and glycoproteins which allows tissues on the external surface of a body (like skin, lung, intestine, etc.) to adhere to underlying tissues. In most cells, an extensive network of intermediate filaments surrounds the nucleus and extends to the cell periphery.

Intermediate filaments constitute a heterogeneous group of cytoskeletal fibers. The most common group is constituted by a protein subunit called vimentin, which gives structural stability to many cells. Keratin, another group of intermediate filaments, is located in ephitelial cells, in particular those which cover the external side of the body, such as skin, hairs and nails. Intermediate filaments of nervous cells are called neurofilaments.

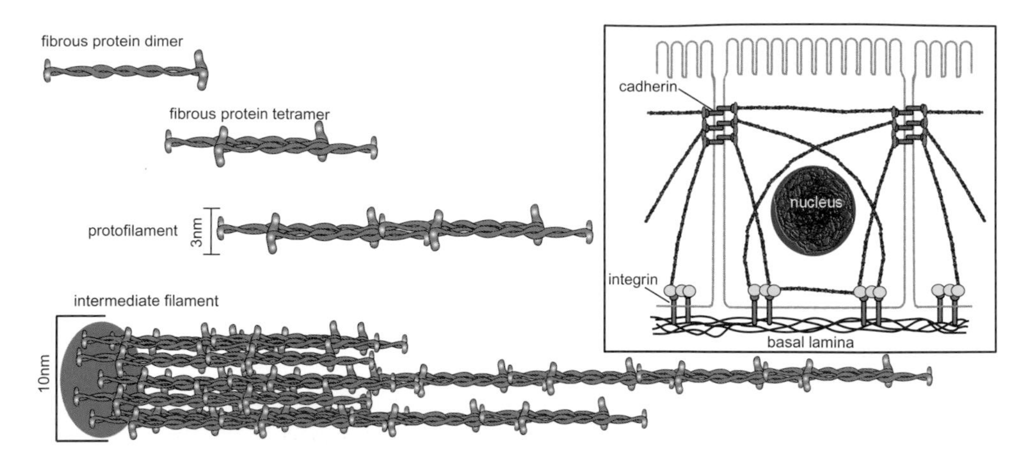

Figure 1.6. Assembly of protein fibers which form intermediate filaments and organization of filaments inside the cell.

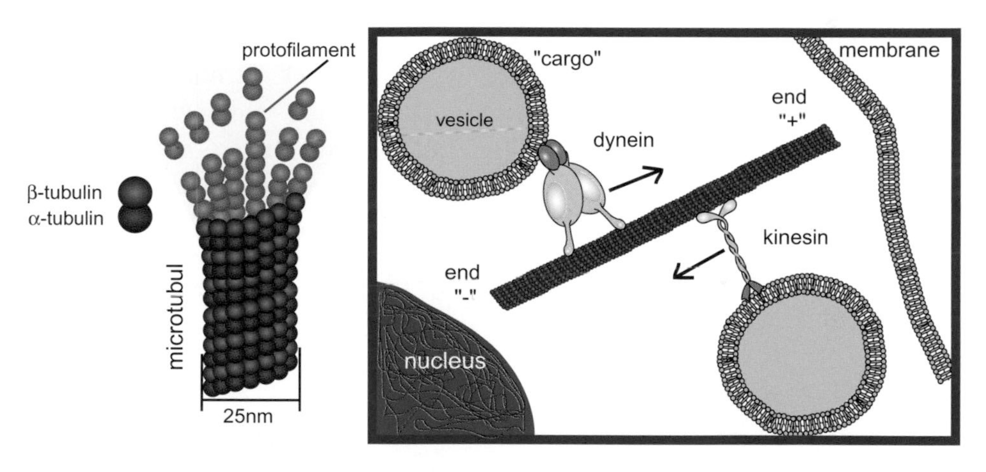

Figure 1.7. Assembly of microtubules and interaction with motor proteins. Polymerization of α and β-tubulin dimers forms protofilaments. Generally, the assembly of 13 protofilaments disposed in a circle forms a microtubule. Two motor proteins interact with microtubules to transport their "cargo": dynein and kinesin. Dynein transports its "cargo" in the "–" direction (torwards the inside of the cell). Kinesin, on the other hand, transports its "cargo" in the "+" direction (towards the outside of the cell).

The microtubules

Microtubules are empty pipes formed in general by 13 protofilaments disposed in a circle (Fig. 1.7). They are extremely rigid, almost 1000 times more than actin filaments. Each protofilament is constituted by two globular proteins: α and β tubulin. Tubulins polymerize using the hydrolysis of guanosine triphosphate (GTP). Microtubules are highly abundant in eukaryotic cells, and in particular inside nerve cells where they represent 10–20% of total protein. Microtubules often originate from a nucleation center located in the central region of the cell, called centrosome, where they irradiate to cell periphery.

Box 1.1. Why we need molecular motors.

With a diffusion coefficient of $10^{-10}\,\mathrm{m}^2\,\mathrm{s}^{-1}$ almost 1000 seconds are necessary for a protein to travel a distance corresponding to the size of an eukaryotic cell with a diameter of 100 μm. This time, which increases with the square root of the distance, will be bigger if the protein must diffuse along the long axon of a nerve cell. To reduce this time, cells need an active form of transport. Actin filaments and microtubules play a role of tracks for the molecular motors: myosins, kynesin and dyneins. These motors consume energy to move at the speed of few μm per second, reducing the transport time to a few dozen seconds over cellular distances.

Like actin filaments, microtubules are continuously polymerized and depolymerized, and like actin filaments they are polarized and the two extremities are not equivalent. Their lifetime varies from 20 seconds to 10 minutes inside animal cells, depending on the cell cycle — microtubules are notably stabilized during cell division when they form the network of “tracks” along which chromosomes are segregated, as discussed in Fig. 1.8. Microtubules are also implicated in different types of cell motion, for instance as components of cilia and flagella. The motion of these specialized filaments are driven by molecular motors, which bend the microtubule bundles located inside cilia and flagella. Microtubules are also responsible for movements of cargos via interaction with “motor” proteins. For

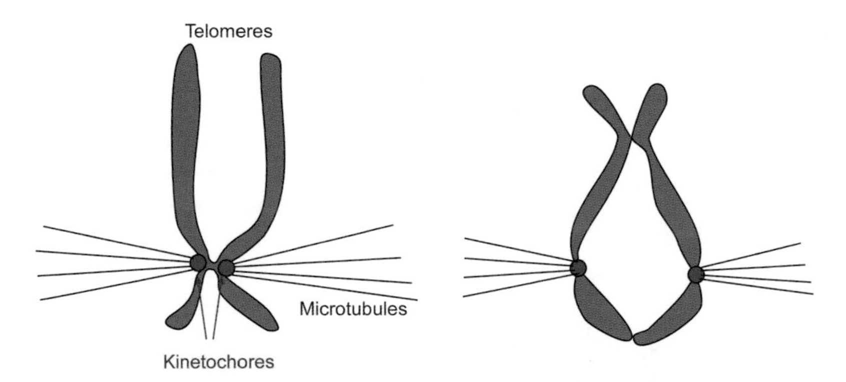

Figure 1.8. Chromosome structure after replication of DNA before cell division. Replicated chromosome is condensed in a characteristic “X” shape. Some DNA sequences, called centromeres, anchor some protein complexes, kinetochores, which allows chromosome to be tracted along microtubules until reaching the opposite cell extremities. Then the cell can divide splitting in two half, in a way that each daugther cell inherits one copy of each chromosome. The telomeres are sequences located at chromosome extremities and they are difficult to replicate.

motors moving along microtubules, in general kinesins move to the plus extremity of microtubules, whereas dyneins move towards the minus extremity (in the direction of the centrosome).

We can cite also the role of microtubules and their motor proteins in chromosomes separation.

In Eukaryotes, mitotic cell division is followed by breakup of the nucleus. After DNA replication, the chromatin DNA structure undergoes "condensation" to form highly compact "mitotic chromosomes", the well-known X-shaped objects (and sometime Y-shaped object) observed in assembled Karyotypes after amniocentesis (Fig. 1.8). At the center of those molecules we can find specific DNA sequences called centromeres, where a protein complex called the kinetochore assembles. The kinetochore acts like an anchoring point for proteins towing the chromosome along microtubules until they reach the cell poles.

After this description of the cell, its compartments and the main elements responsible for its structure, we are going to pass to a more molecular description and we will focus on DNA (deoxyribonucleic acid), the molecule responsible for genetic heridity, and on its use by living organism for protein production.

3 At the molecular scale

3.1 An introduction to the use of genetic information

From bacteria to humans, the genome of an individual is carried by a small number of very long DNA molecules, densely compacted into the cell in the form of chromosomes. Each chromosome consists of a single DNA molecule which encodes genetic information in the form of a linear sequence of four biochemical "letters" (Fig. 1.9) named bases and denoted A, T, G and C according to the initial of the molecule it

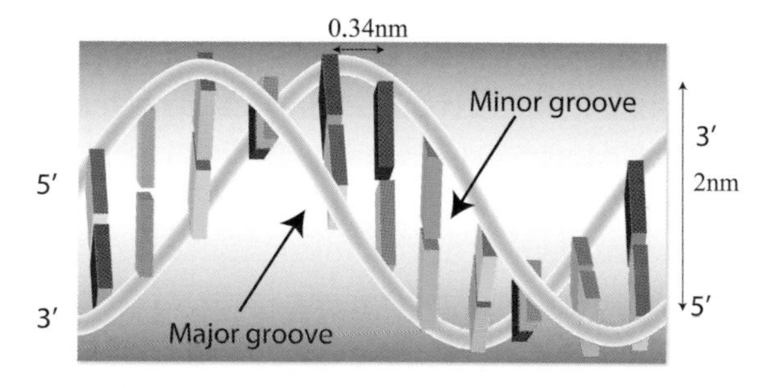

Figure 1.9. Structure of DNA. The double helix is formed by the wrapping of two strands one about the other. Each strand carries the bases of the genetic code, represented as parallelograms. The complementarity between the bases of the two strands ensures the stability of the double helix. Figure adapted from G. Charvin *et al.*, Contemporary Physics, 45, 383–403.

designates: A for adenine, T for thymine, G for guanine and C for cytosine. The functional unit of the genome is the gene. A gene is a unique sequence of 10^2 to 10^4 bases containing the information required for synthesis of a corresponding protein. To synthesize a protein, the DNA of a gene must first be transcribed into RNA, a molecule chemically very similar to DNA and which essentially contains the same information as the gene. The protein which carries out this transcription is named RNA polymerase. The RNA is then taken up by another complex, the ribosome, which synthesizes the protein according to the instructions carried in the RNA. This step is named translation.

Each protein has a precise biochemical function, and sometimes even several, which may be absolutely necessary to the cell. The cell controls the rhythm at which proteins are synthesized by regulating transcription and/or translation. The degradation of proteins and of RNA is also actively regulated by the cell. At each moment in time, the number of molecules of a given protein present in a cell depends on these fluxes of synthesis and degradation. One of the most important reasons for this temporal regulation of the composition of the cell, and therefore of its function, is the organization of cell division. During cell division the DNA of the cell is copied, or replicated, by proteins named DNA polymerases, doubling the number of chromosomes in the cell. The two chromosomes resulting from one original chromosome are then distributed to the poles of the cell before its division into two new cells.

These processes of transcription, translation and replication of the genome form what is named the "central dogma " of molecular biology. This dogma describes the major flows of genetic information within a cell and from one cell to its offspring. Overwhelmingly adopted in the years which followed the discovery of the structure of DNA in 1953, this model has obviously been ameliorated since. However before discussing its ameliorations it is important to know the basis for the system.

The molecules present in the cell are remarkable by the tight relation between form and function. DNA encodes genetic information as a long polymer chain which can wind into a double helix; the mechanical properties of this double helix, which we will see in detail in Chapter 3, are equally essential to its biological function. Similarly proteins are polymeric chains of amino acids which fold into a complex three-dimensional structure dictated by the linear sequence of amino acids and their interactions. They thus become small machines capable of performing work that can be of a purely biochemical nature (for instance by catalyzing chemical reactions), or that can additionally involve mechanical work (for instance during active cargo transport throughout the cell). The cell itself is defined by lipid membranes, which are fluid systems which form both the cell membrane and its internal vesicles; when such vesicle membranes are decorated with the appropriate proteins, their mechanical properties (e.g. curvature) are modified, allowing them to fuse together or bud off new vesicles. In all of these cases, the biochemical activity is related to the physical and chemical structures and properties of the biological molecule. A description of the chemical properties of a biological molecule should therefore be complemented by a description of its mechanical properties as well as its statistical properties.

3.2 DNA

Genetic information is carried by a very long molecule named deoxyribonucleic acid, or DNA. In 1953 the elucidation of the double-helical structure of DNA by James Watson and Francis Crick on the basis of snapshots taken by Rosalind Franklin constituted a fundamental step in our comprehension of the molecular basis of genetic heredity (Fig. 1.9). Each strand of the double helix is a linear organic polymer built of four letters A,T, G or C. The total length of DNA in a chromosome can readily be on the scale of several centimetres and be built of several hundred million bases. The sequence of bases along the length of the polymer constitutes the genetic code.

One can describe DNA as a hierarchical series of structures. The primary structure corresponds to the simple chemical/genetic sequence of the molecule. The secondary structure describe the way in which the two strands of the double helix wrap one about the other. The tertiary structure represents the path drawn in space by the long axis of the double helix: the molecule can for instance wrap about itself or about proteins named histones to form what is denoted chromatin. Chromatin then organises into chromosomes, in a manner which remains only poorly understood.

Primary structure of DNA

As illustrated in Fig. 1.10, the DNA molecule is constituted of two polymeric strands arranged in an antiparallel fashion. The deoxyribonucleotide, also named nucleotide or base, is the unit of repetition of the strand and contains three parts: a nitrogenated aromatic ring (the base), a sugar (deoxyribose) and a phosphate group. As already mentioned there are four nitrogenated bases in DNA, and the base is connected to the sugar by its first carbon atom.

Along each strand, the phosphate group is connected to the fifth carbon atom of one adjacent deoxyribose sugar and to the oxygen linked to the third carbon atom of the other adjacent deoxyribose, forming what is known as a phosphodiester bond. One can thus orient a single DNA strand from the extremity denoted "5′" (terminated by a phosphate group) to its extremity denoted "3′" (terminated by a hydroxyl group). This polarity in the strands of DNA can be essential for proteins

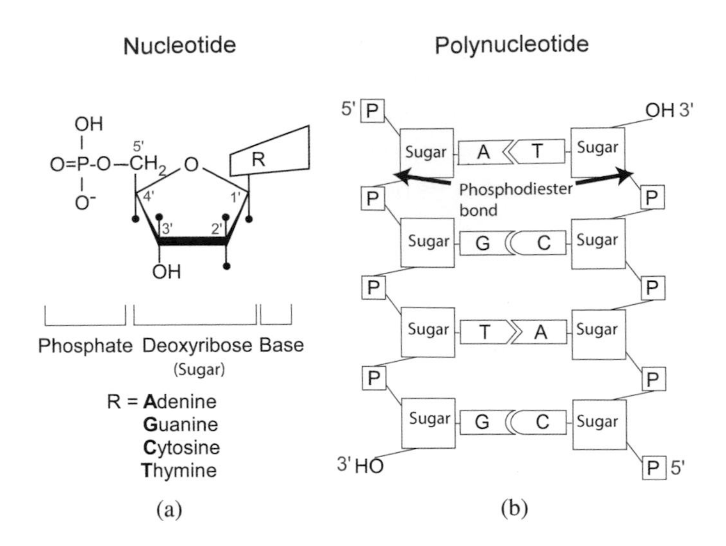

Figure 1.10. Chemical structure of DNA. (a) The unit of repetition of the single-stranded polymer is the nucleotide, composed of a nitrogen-containing aromatic ring, or base (denoted R in the sketch), a sugar molecule, and a phosphate group. (b) The succession of nucleotides forms an alternating sugar-phosphate structure. The two strands of DNA are oriented in an antiparallel fashion from the carbon atom numbered 5′ to the carbon atom numbered 3′ of the deoxyribose sugar.

which will bind to the molecule. It provides directionality to the polymer, which will for instance allow its reading in the correct orientation by proteins which bind to DNA. In the antiparallel double helix, each strand is oriented "upside-down" relative to the other. The chemical bonds ensuring the integrity of a given strand are covalent bonds. At neutral pH, the phosphate group bears a negative charge which will play a role in the physical and biochemical properties of DNA. The distance between successive phosphate groups along the chain is about 0.7 nm.

The bases comprise a planar aromatic cycle which is relatively hydrophobic (Fig. 1.11). The presence of electronegative oxygen atoms on one hand, and of primary (-NH_2) and secondary (-NH) groups which are more or less electropositive, allows G and C bases to interact in stable fashion, and thus to pair together via the formation of three hydrogen bonds, whereas A and T bases are capable of only forming two hydrogen bonds. These paired bases are said to be complementary. Thus the bases on one DNA strand interact and form stable structures with the complementary bases present on the other strand of the DNA molecule.

A DNA chain is polymerized by forming a chemical bond between the 3′-OH end of the chain and the 5′-phosphate of the nucleotide to be added (Fig. 1.12). In this reaction catalyzed by the DNA polymerase enzyme, the nucleotide to be added is initially under the form of a triphosphate (dATP for the adenine base for

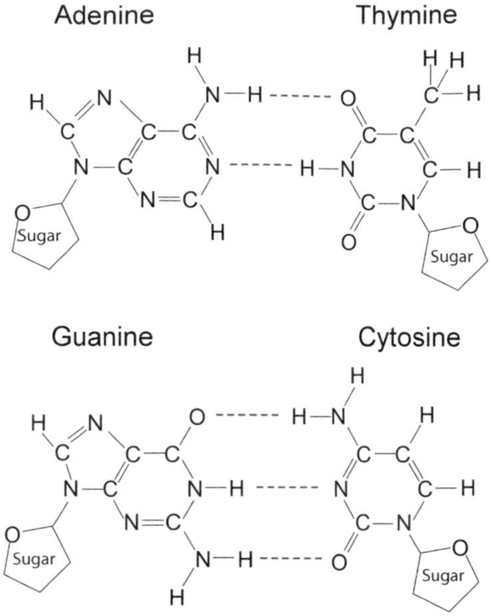

Figure 1.11. Representation of the four bases and their pairing in B-form DNA. The directionality of the hydrogen bonds between guanine and cytosine on one hand, and between adenine and thymine on the other, renders this interaction specific. One notes the asymmetry in the position of the sugar molecules in a pair of bases; this results in a wider ("major") groove and narrower ("minor") groove on the face of the double helix as indicated in Fig. 1.9.

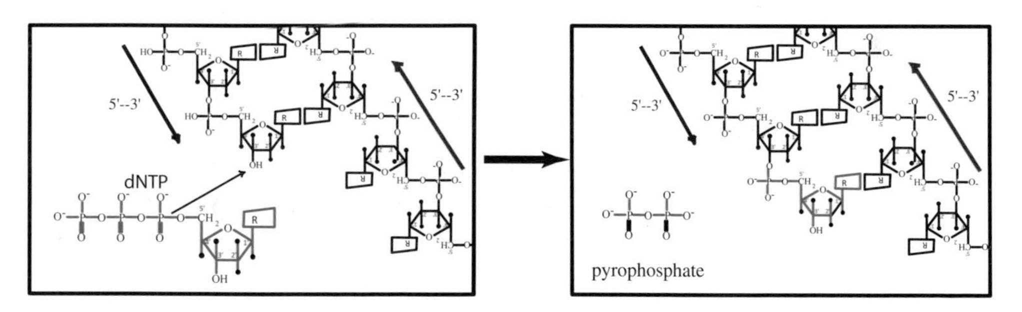

Figure 1.12. During DNA synthesis the DNA polymerase enzyme adds a deoxyribonucleotide (dNTP, with N = A,T, G or C) to the hydroxyl group located at the $3'$ extremity of the nascent chain, releasing a pyrophosphate group from the polymerized NTP. The same DNA polymerase enzyme then moves towards the new extremity of the chain to repeat the same reaction. DNA polymerase thus uses chemical energy to displace itself: it behaves like a motor.

instance), a highly energetic group which stores electrostatic potential energy in the closely-arrayed negative charges carried by the oxygen atoms bound to the phosphate atoms. The polymerization reaction generates a phosphodiester bond with the adjacent nucleotide, releasing two of the three phosphate groups. The new nucleotide can then receive another nucleotide at its 3′-OH end. From a biochemical standpoint it is less risky to proceed in this direction, rather than polymerize the chain at the 5′ phosphate end, precisely because the highly-charged 5′ triphosphate is more energetic and thus less stable than the 3′-OH end. Because the reaction depends on the triphosphate group of the incoming nucleotide, rather than the triphosphate group of the chain laboriously synthesized after numerous reaction cycles, the risk of interrupting polymerization is lowered.

Box 1.2. Primary structure of RNA.

RNA, or ribonucleic acid, is a single-stranded nucleic acid which differs from a single-strand of DNA on two essential points. First of all, the sugar molecule in an RNA base is a ribose rather than a deoxyribose; thus ribose is more electronegative as it carries two hydroxyl groups rather than just one, and this substitution makes the RNA base, or ribonucleotide, less stable than the DNA base, or deoxyribonucleotide. We note that the molecules which store a large portion of the cell's chemical energy are these same ribonucleotides, but under the form of a triphosphate. This is the case of ATP, or adenosine triphosphate, the synthesis of which is discussed in Chapter 4. Second, the thymine base is not present in RNA, which has in its stead a uracil base, denoted U. Uracil pairs specifically with adenine and thus replaces thymine when DNA is converted into RNA by the transcription reaction (see below). As it is essentially single-stranded, RNA is mechanically much more flexible than double-stranded DNA and can fold up on itself when self-pairing is made possible by the sequence of the molecule.

This possibility of folding and self-pairing, coupled with the lower stability of the ribonucleotide, allow for a slow autocatalytic activity at the 3′-OH end of the RNA, using itself as template! Given this potential for slow autonomous replication, it is probable that RNA appeared before DNA during the course of evolution.

In the cell, RNA is synthesized during a reaction termed "transcription" which is carried out by the RNA polymerase enzyme (discussed in more detail in Chapter 4). The basic principle of the synthesis reaction is close to that for DNA (Fig. 1.12), but in which the deoxyribonucleotide (dNTP) is replaced by the ribonucleotide (NTP, which N = A, U, G, or C) while the template remains the DNA. RNA can be found in many forms: by itself as the messenger RNA (mRNA) used to code for protein synthesis; covalently coupled to an amino acid as transfer RNA (tRNA) used during translation of the genetic code into protein, or integrated into a large protein-nucleic acid complex as is the case for the ribosomal RNA (rRNA) which is an integral part of the ribosome, the ultimate protein-synthesizing machine responsible for translation.

Secondary structures of DNA: the double helices

When the two polymeric chains which constitute DNA wrap one about the other, the hydrogen bonds formed between the complementary bases of the two strands stabilize the double helix. The double helix is further stabilized by the topological wrapping of the two strands about each other. In addition, the relatively hydrophobic nature of the nitrogenated bases means that they shy away from aqueous solution, and given their planar structure they tend to stack atop one another in a manner reminiscent of a stack of plates, thus also contributing to the stability and the rigidity of the double helix. The negative charge of the phosphate groups also rigidifies DNA via the electrostatic repulsion of adjacent bases along a single strand.

The standard (or "canonical") form of DNA in solution is named B-DNA, and as shown by Watson and Crick it is a right-handed double-helical structure about 2.4 nm wide and with a helical pitch of 10.5 base-pairs per turn. The vertical rise between base-pairs along the direction of the helix axis of the molecule is about 0.34 nm.

The different interactions which stabilize the double-helical form of DNA also confer it with very high local stiffness: the persistence length ξ of B-DNA, i.e. the distance over which the orientation of the helix axis persists despite thermal agitation) is $\xi = 50$ nm (see Chapter 3), or approximately 150 base-pairs.

The structure and stiffness of the double helix is readily appreciated: an RNA molecule, structurally related to a single strand of DNA, can adopt very complex and abrupt stuctures in which the molecule's path rapidly changes orientation on the scale of only a base or two, such as in so-called "hairpin structures" based on inverted repeats of sequence for instance.

As a function of the experimental conditions (solvent, salts, force, torque) numerous additional DNA structures of DNA have been observed, some with helicity close to the canonical form, others with inverted helical pitch (such as for left-handed Z-DNA, which also has a pitch of approximately ten per turn) or an over-elongated form (such as so-called "stretched", or S-DNA, which is a flattened double helix with a pitch of 23 bases/turn). If the great majority of DNA in a cell is B-form, a portion of it could transiently find itself in other forms, in particular when these are stabilized by interactions with DNA-binding proteins.

More generally, the topological implications of the double helical structure remain both deep and subtle. For instance as originally pointed out by Watson and Crick, if the structure of the double helix provided a structural basis for the redundant storage of information, it posed an important topological dilemma during replication. How must the initial DNA be separated so as to be replicated into two independent, non-intertwined, DNA molecules?

Tertiary structure of DNA: supercoiling and packaging

DNA is also remarkably rigid from a torsional standpoint, with a torsional persistence length of about 85 nm or nearly 250 bp. As a double helix, it can furthermore be twisted and supercoiled according to a simple topological formalism known as the linking number description. This would not be possible with a polymer built of only a single chain of covalent bonds for instance. *In vivo* DNA is slighly underwound, or negatively supercoiled, as if one had unwound a bit of double helix by

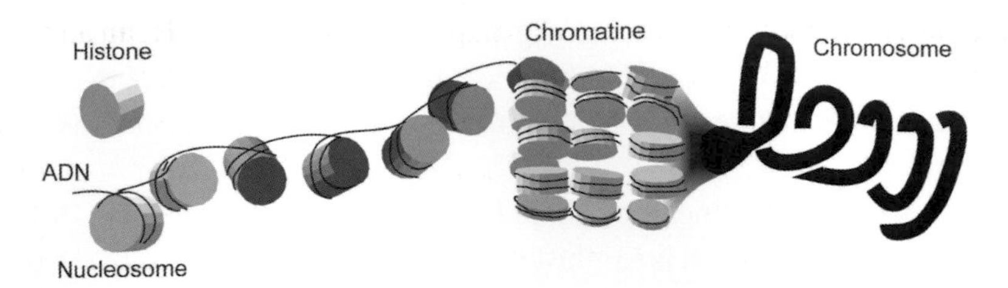

Figure 1.13. Hierarchical packaging of DNA in eukaryotes. DNA wraps in a solenoidal fashion about histone proteins to form the nucleosome. A succession of nucleosomes can fold together to form chromatin, a more complex filament with a structure which remains still insufficiently understood. Chromatin in turn folds up on itself to form the chromosome. Each stage makes it possible to progressively package DNA into a smaller and smaller volume.

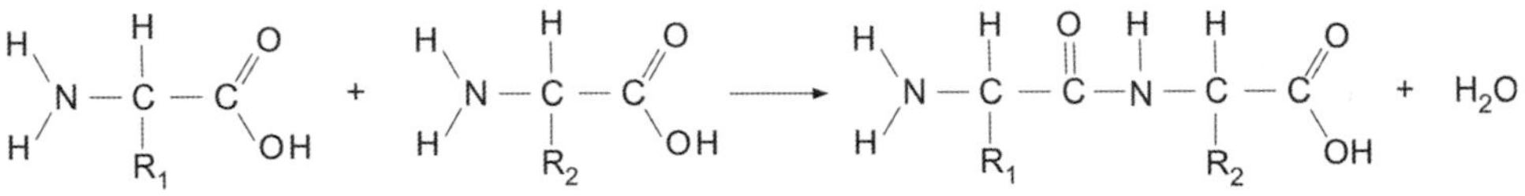

$$\mathrm{H_2N-CH(R_1)-COOH} + \mathrm{H_2N-CH(R_2)-COOH} \longrightarrow \mathrm{H_2N-CH(R_1)-CO-NH-CH(R_2)-COOH} + \mathrm{H_2O}$$

Figure 1.14. Chemical structure of an amino acid. The amino acid is comprised of a central carbon surrounded by four chemical groups of which three are constant and one, the radical, is variable. Two amino acids can be connected by a peptide bond. A chain containing multiple amino acids is named polypeptide. A protein can be constituted of one or several polypeptides.

untwisting the molecule at one end. This torsional constraint can be stored either as unwinding of DNA, but also by wrapping DNA around small cylindrically-shaped packaging proteins known as histones (Fig. 1.13). DNA wrapped about a histone protein forms the so-called "nucleosome", and nucleosomes are found in succession along DNA like "beads on a string". Certain regulatory proteins modifiy histone proteins to change the degree of DNA packaging in the nucleosome and the higher-order chromatin fibre which builds up chromosomes. In our cells, such packaging allows each cell to store approximately 2 metres of linear DNA material into a cell nucleus of about 5 μm diameter. This is to be compared with the volume occupied by the same amout of DNA absent any packaging and as dictated by polymer mechanics, namely a 200 μm radius sphere!

3.3 Proteins

Primary structure

Proteins are polymers for which the unit of repetition is an amino acid (Fig. 1.14). An amino acid is constructed around a carbon atom, denoted C, and connected by simple covalent bonds to four chemical groups: a hydrogen atom H, an amine group NH_2, a carboxylic acid COOH, and a radical denoted R. In nature there are twenty different radicals, the chemical properties of which are used in a combinatorial fashion to create proteins. Amino acids polymerize by forming a peptide bond CONH between the carobxylic acid of the chain's last amino acid and the amine group of the next amino acid to be added. Proteins are thus said to polymerize from N-terminus to C-terminus, where the "N-terminus" corresponds to the amine group of the first monomer in the chain and the "C-terminus" corresponds to the carboxylic acid of the last monomer of the chain.

Secondary structure

If strong covalent bonds connect amino acids to each other, these can also interact with each other via the chemical groups of their radicals, also known as lateral chains. These lateral chains can roughly be distinguished by their predominant chemical property: positive charge, negative charge, aromatic groups or other non-polar groups for instance. The interactions between amino acids along the chain can be sufficiently strong to fold up into helices (known as α helices) or flat sheet-like structures (known as β sheets). Proteins thus fold up into small structured domains ranging from tens to a hundred amino acids, and these domains can interact with each other to structure the proteins and generate its function.

Tertiary structure

Proteins which act chemically on their substrates possess a catalytic region known as the active site, or active center, in which the chemical reaction takes place. The chemical activity depends on the chemical micro-environment set up by the lateral chains which surround and indeed form the active site; it also depends on the orientation and mechanical strain of the substrate in this micro-environment. The local steric structure of the site, down to the atomic-scale description of the positions of the atoms which form it, is also key to understanding which molecules can bind to the site and undergo catalysis — i.e. what makes the site selective to only certain specific substrates. There also exist many examples of active sites situated at the interface between two structural domains, making the presence of the active site dependent on the global structure of the protein, which may itself be regulated by other proteins of biomolecules. Finally, catalysis at the active center can not only be modulated by protein structure, but it can also itself drive changes in protein structure and allow proteins to "cycle" through a series of structures in a manner not so different from a small engine.

3.4 Transcription of DNA into RNA

The first step on the path to protein synthesis based on the information carried by a gene is termed transcription. The RNA polymerase enzyme reads the DNA sequence of a gene and re-writes it in the form of a complementary RNA strand. To localize the beginning of the gene, the RNA polymerase looks for a sequence known as a promoter — similar to the capital letter at the beginning of a sentence (Fig. 1.15) which allows a reader to distinguish separate sentences. The DNA sequence which makes up a promoter has specific features which must be respected to make an efficient promoter, but variations of sequence also make it possible to tune the efficiency of the promoter. The physics of promoter search of this sequence target in the cell will be discussed in Chapter 9. Upon promoter binding, RNA polymerase unwinds a bit over a full turn of the double helix, or about 12 base-pairs for the *E. coli* RNA polymerase. The RNA polymerase thus unwinds DNA a bit beyond the first base to be transcribed and positions the first base to be read on the so-called "template" strand at the active center.

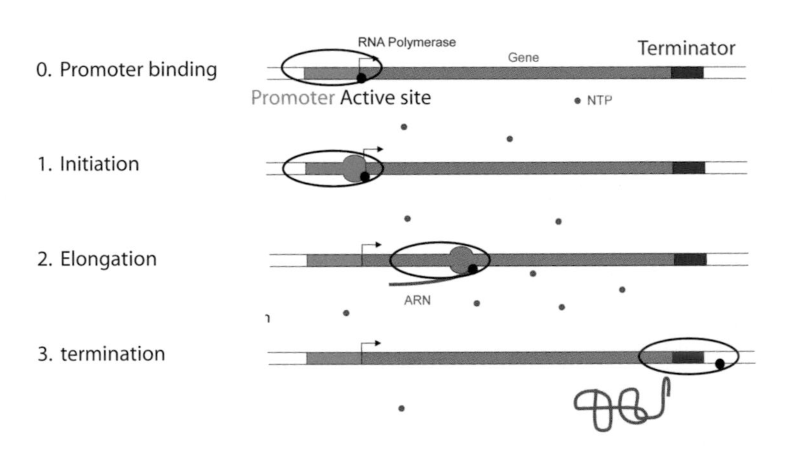

Figure 1.15. Transcription of DNA. (0) RNA polymerase identifies the beginning of a gene to be transcribed by binding to the promoter sequence. (1) At transcription initiation, RNA polymerase unwinds DNA at the promoter site so as to accurately read the chemical structure of the first base to be transcribed. (2) During elongation, RNA polymerase starts reading the DNA and escapes from the promoter, reading the successive bases of the bottom strand in the $3' \rightarrow 5'$ direction to synthesize complementary RNA in the $5' \rightarrow 3'$ direction. (3) At the end of the gene, an RNA polymerase may encounter a termination sequence, which allows it to free both the RNA and the RNA polymerase from the DNA.

Thus the promoter serves as a starting point for gene transcription. The template strand is the "bottom" strand when the DNA is drawn horizontally and the active center moves rightward relative to the DNA. As RNA is complementary to the bottom strand, it has the same sequence as the standard ($5' \rightarrow 3'$) sequence of a gene. Here again the chemical orientation of DNA is essential; without it polymerase would be like a reader who would start at a capital letter of a sentence but then be just as likely to read backwards (the prior sentence) as forwards (the new sentence). Polymerization of the new nucleic acid takes place, as always, in the $5' \rightarrow 3'$ direction. At each base of the template strand, a nucleotide triphosphate diffuses into the catalytic site, which retains only the nucleotide triphosphate complementary to the base present in the site (the same base-pairing rules apply as for DNA except that in RNA a U is found instead of T). A non-complementary nucleotide which may diffuse in is not retained because the mismatch between the non-complementary bases prevents the nucleotide from "fitting" correctly in the active center and inhibits polymerization of the incorrect nucleotide. The complementary nucleotide binds with its $5'$-triphosphate near the $3'$OH end of the growing RNA chain. The chemistry of the active site facilitates nucleophilic attack by the $3'$OH end on the first phosphate of the triphosphate group, binding the incoming nucleotide onto the $3'$ end of the growing chain and releasing inorganic pyrophosphate (PP_i). At the end of this reaction, the RNA polymerase advances along the template DNA strand by one base and repeats this nucleotide addition cycle, generating an RNA complementary to the template strand. Thus in a way RNA polymerase uses chemical energy to displace itself: it is therefore a molecular motor (see Chapter 4).

Arrived at the end of the gene, RNA polymerase dissociates from the DNA and the new messenger RNA molecule is released. If the gene is the target of multiple successive rounds of transcription, an elevated number of messenger RNA molecules will be available for the cell; such a gene is said to be strongly transcribed.

Transcription can be regulated at each of these stages. The binding of RNA polymerase to the promoter can be stimulated by an "activator" protein, for instance which will bind just upstream of a promoter to then tether RNA polmerase in the vicinity of the promoter and thus help "recruit" it to the promoter. Other proteins can act as repressors of transcription, for instance by binding on, or near the promoter and preventing RNA polymerase from binding to the promoter, or escaping from the promoter to carry out full gene transcription.

The interactions between genes and their products (e.g. proteins) can be represented under network form where a gene represents a node and the molecular interactions between the product of that gene and its target represent the connections.

These connections must thus reflect the fact that interactions can be repressive or activating. Extensive efforts are under way to map these complex transcription networks to try to identify and interpret key motifs and find simple laws that govern these complex systems.

3.5 Translation of RNA into protein

Messenger RNA is taken up by the ribosome for translation into protein (Fig. 1.16). The ribosome is a hybrid system composed of several RNA molecules (three in bacteria, four in eukaryotes) and more than 50 proteins or polypeptides. The ribosome progressively reads messenger RNA by successive, 3-base blocks named codons. To each codon corresponds an amino acid; this correspondence is materialized by a so-called transfer RNA, denoted tRNA. tRNA is another hybrid system composed of an 80-base RNA containing the anti-codon sequence (complementary to the codon) and to which is attached at the 3′-OH end the C-terminus of the corresponding amino acid. For each codon along the mRNA, the ribosome retains the appropriate tRNA in its active site by the complementarity between the bases of the codon and the anti-codon. The ribosome also retains the C-terminal end of the protein undergoing synthesis, carried by the tRNA complementary to the codon just prior that currently being read. At the active site the C-terminal extremity of the nascent polypeptide is transfered to the N-terminus of the amino acid carried by the new

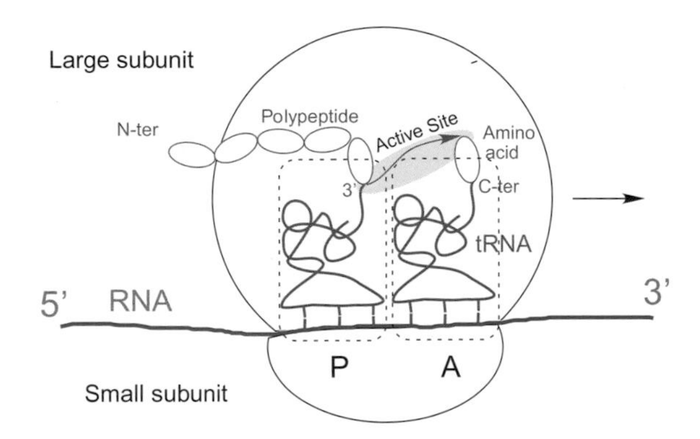

Figure 1.16. Translation of RNA. The ribosome reads RNA codon by codon. Facing each codon is located a tRNA bearing the complementary anti-codon and the corresponding amino acid. The amino acid is connected via its carboxy group to the tRNA. The ribosome can accomodate two tRNAs; one tRNA bears the polypeptide chain synthesized so far from this RNA (it occupies the site denoted "P"; the second tRNA is just adjacent and carries the amino acid specified by the next codon for addition to the polypeptide (it occupies the site denoted "A"). The active site of the ribosome transfers the C-terminal end of the polypeptide from the tRNA in site P to the N-terminus of the amino acid carried by the tRNA in site A. The ribosome advances by one codon along the RNA and the tRNA previously in the A site now finds itself in the P site, so that a new cycle of catalysis may take place.

adjacent tRNA. The depleted tRNA from the prior cycle is released and the ribosome advances by one codon along the messenger RNA before beginning the cycle anew. The ribosome thus also in a way behaves like a molecular motor, moving along RNA to read it and synthesize protein. The end of the gene is indicated by a so-called "stop" codon which causes dissociation of the RNA/ribosome/protein complex. In bacteria, transcription and translation can take place in parallel, with multiple ribosomes translating a single RNA as it is being transcribed by an RNA polymerase.

Remarkably, it has been observed that messenger RNAs can be found which regulate their own translation as a function of the metabolic conditions of the cell. In these RNAs, the so-called ribosomal binding sequence (RBS) which allows for initial binding of the ribosome to the RNA can be made inaccessible if it is tucked away in a folded portion of the RNA. So-called riboswitch sequences, typically localized at the beginning of genes, will thus fold up on themselves by self-complementarity only in the presence of a specific small metabolite (such as an amino acid for instance) and sequester the RBS, preventing translation. Thus a riboswitch which folds up in the presence of the amino acid lysine is present in the messenger RNA of different proteins involved in the lysine metabolic pathway. If the metabolite is present in sufficient concentration, the riboswitch is folded and translation of proteins required to synthesize the metabolite is shut off. If the metabolite is absent, the riboswitch is not folded and the accessible RBS allows translation, regenerating the metabolic stock of the cell.

Thus gene expression can be regulated at numerous stages: transcription, translation, and also post-translational modification of proteins as it takes place in the Golgi apparatus, an element of the cell discussed earlier.

3.6 Replication of DNA

Just as for the processes of transcription and translation, DNA duplication, or replication, is actively regulated by the cell. It is one of the earliest steps in the complex process of cell division, and initiation of replication is under tight temporal control by the cell. Replication is semi-conservative, meaning that each of the two DNA molecules generated by replication will possess one of the strands of the original double helix. Thus each of the two strands of the original molecule is read and copied to generate a new double-strand DNA molecule. A small biochemical factory named the "replisome" is responsible for this task, and it contains at least two DNA polymerase proteins, one for each strand to be replicated. A recent measurement has measured the average number of polymerases in this factory as three for *E. coli*.

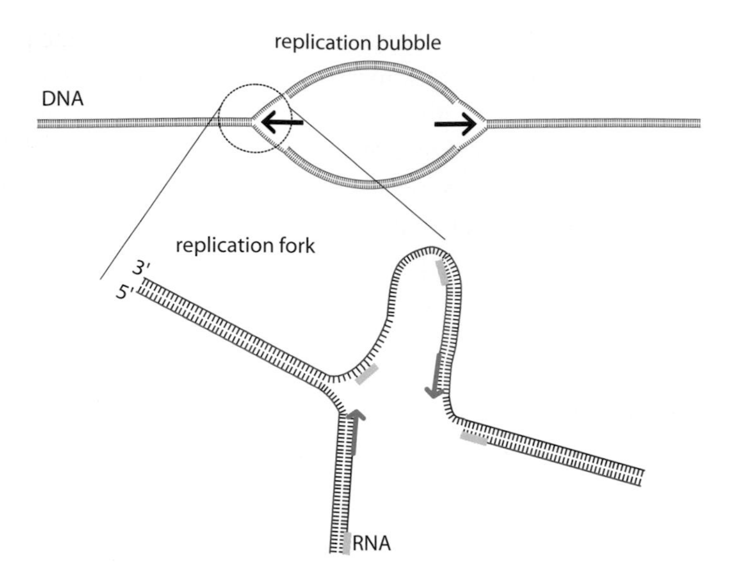

Figure 1.17. DNA replication. Replication begins by DNA opening at a sequence termed an origin of replication. At the two edges of the open region are localized the replication forks. A replisome assembles at each fork, bringing together a number of proteins (not depicted) required to unwind DNA ahead of the DNA polymerases which will copy it. DNA polymerase requires an RNA primer (in green) to begin synthesis. As DNA polymerase can only polymerize DNA in the $5' \rightarrow 3'$ direction, and because the two strands of DNA are antiparallel, the two strands at the fork are not replicated in exactly the same manner. Consider the right-most fork; for the bottom strand oriented $3' \rightarrow 5'$ replication can proceed towards the right in a continuous fashion. The top strand forms a large loop so that a DNA polymerase can replicated it $5' \rightarrow 3'$ by moving towards the fork. Thus on the top strand, a loop is repeatedly formed and replicated, in the direction contrary to overall progression of the fork, to generate small discontinuous segments of DNA known as Okazaki fragments.

Replication of DNA begins at specific sites, termed origins of replication (Fig. 1.17). At these sites, at the appropriate moment, DNA is unwound by proteins which initiate replication, forming an unwound "bubble" flanked by two "forks". Each fork is stabilized by the assembly and action of a replisome, and progression of the fork requires DNA unwinding downstream of the fork to replicate the DNA and yield unlinked DNA molecules at the end of replication.

Assembled at the origin, the replisome contains several proteins including a primase which synthesizes on each strand of the fork a small complementary RNA fragment which serves as a primer for DNA polymerase which is ultimately responsible for DNA replication. Biochemically, the synthesis of DNA is similar to that of RNA, and consists in polymerizing at the 3′-OH end of the nascent strand the correct base complementary to the template strand and held transiently in the active site of the polymerase. The replisome also contains, in addition to the primase and three polymerases, a helicase which unwinds the DNA and thus materializes the fork itself, two "β-clamp" proteins which form a ring about single-strand DNA and serve as a loading platform to help stabilize DNA polymerase on the DNA, and a protein named tau which serves as a molecular scaffold onto which the different components of the replisome are assembled.

As the two strands of DNA are antiparallel and entwined, several constraints arise during replication. As each strand of DNA at the fork must be read in the $3' \rightarrow 5'$ direction for polymerization of the complementary strand to take place in the obligatory $5' \rightarrow 3'$ direction, the two strands are not replicated in an identical manner. Instead, one strand is replicated in a continuous fashion, the other strand is replicated in a discontinuous fashion by the synthesis of small segments known as "Okazaki fragments".

As already mentioned, bacteria often posess circular genomes — and this causes a problem during replication. For instance for *E. coli* with a circular genome of about 4×10^6 bases, replication generates two new genomes with potentially 4×10^5 turns one about the other, i.e. the number of turns in the double helix of the initial genome. For cell division to complete successfully these two genomes must be unlinked, as otherwise chromosome segregation (or cytokinesis) just prior to cell division will fail and the cell will not be able to divide — indeed neither the cytokinesis machinery nor the cell septation machinery can generate forces large enough to break DNA. Topoisomerase enzymes (Fig. 1.18) are capable of carrying out complex reactions in which DNA is unlinked and torsionally relaxed so as to continuously manage the topological constraints that DNA is subjected to during the cell cycle. In replication, a key topoisomerase carries out unlinking of replicated DNA by binding to crossings between tangled DNAs, cutting one DNA and then passing the intact DNA through the gate before resealing the gate DNA. It is not surprising that given the central nature of DNA supercoiling and linking *in vivo*, topoisomerase enzymes are important targets for both antibiotics (in which bacterial growth is inhibited by specifically targeting topoisomerases that exist only in bacteria) or antitumorals (in which tumor growth is inhibited by blocking topoisomerase activity preferentially in rapidly dividing, i.e. cancer, cells).

Because eukaryotic genomes are typically linear and not circular like bacterial genomes, they posess extremities, and the ends of chromosomes are slightly unstable to replication — i.e. they can be incompletely replicated. Thus chromosome extremities have developed specific repeated sequences, known as telomeric repeats, that act as a buffer to imperfections in replication of the extremities of chromosomes; if telomeric repeat sequences are truncated accidentally, they contain no intrinsic information. However over many cycles of replication they may become

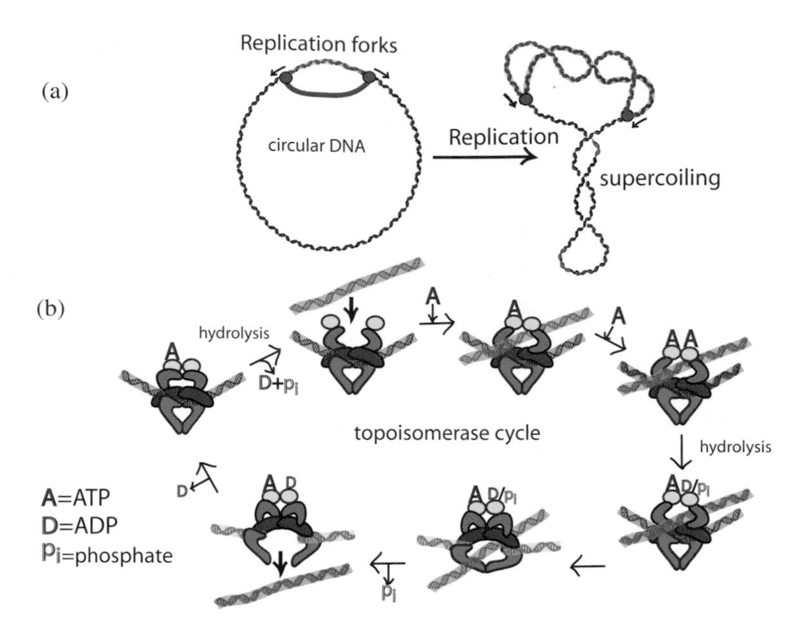

Figure 1.18. (a) During replication of a circular bacterial genome, two replication forks work in opposite directions and should produce entangled, newly-synthesized DNA upstream of the forks and/or supercoiled DNA downstream of the forks. This is because DNA polymerases tracks the DNA groove as it polymerizes. Thus in principle if the forks cannot rotate relative to each other there should be one link and one supercoil for every 10.5 bases replcated. These links must all be removed for replicated DNA to be separated and dispatched, one chromosome to each daughter cell. (b) To resolve the topological problem of tangled DNA after replication, the topoisomerase class of enzymes come into play. Enzymes from this family can for instance bind to one DNA segment (blue) and then a second DNA segment (green) and then transport one DNA through another by transiently cutting the former (there the green DNA is passed through the blue DNA) and then resealing the nick. This reaction requires an energy input, which is the ATP hydrolyzed by the enzyme during its reaction cycle.

progressively shortened, and a reduction in the number of repeats — the length of the buffer — could be related to diseases such as cancer or even ageing. Thus proper control of replication, from beginning to end, is essential to the proper maintenance of a genome.

3.7 New perspectives on the regulation of gene expression

Despite the great success of the "central dogma" during the fifty years that followed the discovery of DNA, a vast amount of complementary and essential information has come to flesh out this model. We have so far here only described "classical" gene regulation, in which for instance a protein will modulate the expression of a gene by binding to its promoter and assisting or hindering RNA polymerase binding to the promoter. More recently the role of DNA packaging itself has been found to be key to understanding how genes are regulated in higher organisms.

As mentioned previously, DNA is packaged in the eukaryotic cell by binding it to proteins known as histones. The histone is a small protein cylinder 5 nm tall and about 8 nm wide. The circumference of the cylinder bears a positive charge that forms a spiral at the surface of the histone; a length of about 50 nm of DNA wraps twice around the histone surface like a solenoid (about 150 bp of DNA, Fig. 1.13). The DNA-histone complex is named nucleosome. Since DNA is mechanically rather stiff, here in particular to bending, the electrostatic interaction between histone and DNA is essential to stabilizing the nucleosome. The stabilization has been confirmed by the role of histone charge modification in the regulation of gene expression.

Indeed, the packaging of DNA around nucleosomes, and then into more elaborate structures (chromatin, chromatin fibres, chromosomes) also makes it possible to regulate genetic activity. It has historically been observed that chromosomes could be stained using appropriate dyes to identify dense and compact regions of chromosomes which could be stained and were named heterochromatin, in contrast to less-compact and less-well stained zones known as euchromatin. Euchromatin is associated with strong transcriptional activity, and heterochromatin is associated with repression of transcriptional activity, also known as gene silencing. If our understanding of the formation of heterochromatin or euchromatin is far from complete, great progress in this field has identified a new code, parallel to the genetic code and named epigenetic.

Very schemactically, if a gene codes for a protein then its epigenetic status determines whether it is transcribed or repressed. This epigenetic status can come from chemical modification of the DNA, most notably by the addition of a methyl group ($-CH_3$) to the cytosine base in the sequence GC; not surprisingly these marks are frequent in so-called GC-repeat regions. These repetitions can be found in the promoters of genes, and their degree of methylation seems correlated to the transcriptional activity of the gene. The epigenetic status of a gene can also be influenced by the chemical modification of histones about which DNA is packaged. Indeed numerous lateral chains of the histone are positively charged, which helps stabilize its interactions with DNA. These lateral chains can also be chemically modified by other proteins; for instance they can be acetylated (by the addition of a $-COCH_3$) or methylated ($-CH_3$) to reduce their positive charge. One could imagine these modifications would alter the packaging of DNA about the histone and improve the efficiency of transcription, eventually asisting in converting heterochromatin to euchromatin. Finally, small RNA molecules complementary to the messenger RNA of genes have also been found to be capable of inhibiting gene expression by hybridizing to the messenger RNA and preventing translation by the ribosome, or also by targeting the gene for repression and silencing. The regulation of gene expression and gene duplication still holds many surprises.

Bibliography

[1] Bruce Alberts, Alexander Johnson, Peter Walter, Julion Lewis, Martin Raff, Keith Roberts. *Molecular Biology of the Cell*, Garland Publishing Inc. (2007).

[2] B. Lemin, *Genes IX*, Jones and Bartlett Learning (2007).

[3] Thomas D. Pollard and William C. Earnshaw, *Cell Biology*, 2nd Edition, Saunder (2007).

2

Fluorescence microscopy for biological imaging

Maxime DAHAN, *CNRS researcher, Laboratoire Physico-Chimie Curie Institut Curie and CNRS, Paris.*

1 Introduction

Optical microscopy has long been an essential tool for the exploration of the living world. As early as 1665, the English physicist Robert Hooke used a telescope to observe thin slices of wood and observed that they were composed of small structures. He had just discovered in plants the existence of cells, the elementary blocks of the animal and vegetal world. At the same period, the Dutch scientist Antoni van Leeuwenhoek (1632–1723) improved the quality of optical components and built the first microscopes. Thanks to his instruments, he made many seminal discoveries, including the existence of single-celled organisms such as bacteria and yeasts, but also sperm and red blood cells.

Since these early works, microscopes have been widely gaining in resolution and sensitivity. In the 19th century, the new understanding of the laws of optics led to large improvements in the design of optical instruments. In particular, Ernst Abbe determined the fundamental limit of resolution in an image. These conceptual advances coincided with the growth of the optical industry and the emergence of companies such as Carl Zeiss and Leica that will market and make these instruments available to many scientists. Nowadays, optical imaging is used daily to understand the functions of our cells, both in fundamental and applied research and for medical diagnosis. Microscopes are no longer mere optical instruments consisting of a set of lenses and mirrors. They include various light sources to excite the samples, opto-electronic devices to detect and record light signals, and they make great use of the power of modern computers to process and analyze these data. In addition, thanks to the considerable advances in molecular and cellular biology, it is now possible to selectively modify the molecular actors of our cells and to couple them with optical markers. The combination of optical, chemical, biological and computational techniques provide modern microscopes with extraordinary sensitivity and resolution, allowing measurements down to the real-time recording of individual proteins in a living environment.

2 Fluorescence

2.1 Principle

When performing optical measurements under a microscope, a contrast method is required for distinguishing the object of interest within the sample. In practice, the most common methods, and as we shall see the most sensitive, are based on a common physical mechanism: the fluorescence. An object (atom, molecule, crystal, etc.) is called fluorescent when, once brought in an excited by light at wavelength λ_{ex}, it can decay by emitting light at wavelength λ_{em}. Since the energy emitted is lower than that absorbed, λ_{em} is longer than λ_{ex}. In the case, frequent in biology, where excitation occurs in the visible range (λ_{ex} between 400 and 700 nm), it results in a spectral shift of the emission towards the red. The typical duration of an excitation/emission cycle is a few nanoseconds. In other words, a fluorescent system may emit a few hundred million photons per second. Knowing that the photodiodes or CCD cameras today allow detection of a single photon, we find here a first explanation for the high sensitivity of fluorescence measurements.

The main properties of a fluorescent system can be understood using a simple two-level model. In this model, a molecule exists in two states: f the ground state and the excited state e in which it is transferred to after absorption of a photon of wavelength λ (in this model we do not distinguish λ_{em} from λ_{ex}) (see Fig. 2.1). To describe the excitation process, we consider that there is an absorption rate k_a, proportional to the incident photon flux per unit surface. The coefficient of proportionality σ_a, called the absorption cross section, has the dimension of a surface and reflects the ability of the molecule to absorb photons at this wavelength. The photon flux is generally written as $I/(hc/\lambda)$, where I is the flux intensity of the excitation beam (in W/m^2) and hc/λ represents the energy of a photon (h is Planck's

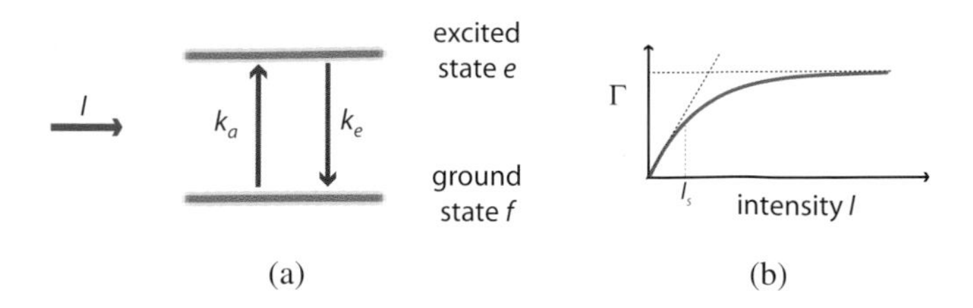

Figure 2.1. (a) Two-level system excited with light intensity I. k_a is the absorption rate and k_e is the decay rate. (b) Fluorescence emission rate Γ as a function of the intensity I. It scales linearly at low intensity and saturates when $I \gg I_S$.

constant and c is the speed of the light). Once excited, the molecule can decay by emitting fluorescence light with a rate k_e. $T_e = 1/k_e$ represents the average time spent in excited state before the emission of a photon and is generally on the order of a few nanoseconds.

The probabilities P_f and P_e of being in the ground state or the excited state satisfy the following equations:

$$\frac{dP_f}{dt} = -k_a P_f + k_e P_e \tag{2.1}$$

$$P_f + P_e = 1 \tag{2.2}$$

At steady state, $P_e^{\mathrm{st}} = k_a/(k_a+k_e)$. Given that the fluorescence rate Γ of the molecule is equal to $k_e P_e^{\mathrm{st}}$, one deduces that:

$$\Gamma = k_e \frac{\sigma_a I/(hc/\lambda)}{\sigma_a I/(hc/\lambda) + k_e} \tag{2.3}$$

more commonly written as:

$$\Gamma = \frac{1}{T_e} \frac{I/I_s}{I/I_s + 1} \tag{2.4}$$

$I_s = k_e hc/\sigma_a \lambda$, called the saturation intensity, is an intrinsic property of each fluorescent compound. At low intensity ($I \ll I_s$), the fluorescence rate scales linearly with intensity $\Gamma \simeq \sigma_a I/(hc/\lambda)$. This is due to the fact that the average time to absorb a photon $1/k_a$ is much longer than the decay time T_e. Thus, the emission rate is determined by the excitation process. In the opposite limit ($I \gg I_s$), called the saturation limit, T_e is much larger than $1/k_a$ and determines the fluorescence emission rate $\Gamma \simeq 1/T_e$.

2.2 Fluorescent probes for biology

With some important exceptions, biological molecules of interest, such as nucleic acids or proteins, have no fluorescence properties in the visible range. Therefore they need to be coupled to fluorescent probes in order to be detected and to study their properties. Labeling and targeting techniques are extremely diverse and primarily depend on the chemical and biological characteristics of the probes.

Organic molecules and labeling techniques

The most common fluorescent probes, which have been used since the early days of microscopy, are organic dyes. These are usually small synthetic molecules whose optical properties, such as the absorption or emission wavelength, or emissivity can be adjusted by modifying their chemical structure and carbon chain length. These probes possess also reactive groups, such as amines or carboxylic acids, so that they can be covalently bound to other molecules. In some cases, it is possible to attach the probe directly to the molecule of interest. However, it is often necessary to go by an indirect labeling method in which the probe is coupled to a molecule which is then used to target the molecule of interest. This is the case for example, in immuno-labelling, where the fluorescent probe is attached to an antibody capable of binding to a specific antigen. In this case, the attachment of the probe to the target is not covalent but the result of a chemical affinity and is governed by the rules of chemical equilibrium. Labeling techniques, direct or indirect, by organic fluorescent probes are extremely common in biology and medicine, and there are hundreds of fluorescent molecules differing not only in their optical characteristics but also in their sensivity to physico-chemical parameters of their environment such as pH, calcium ion concentration or the electrical potential. However, they have an important limitation related to the phenomenon of photodestruction. After each excitation process, instead of relaxing by emitting a fluorescence photon, the molecule has a small, but finite, probability of inducing chemical reactions leading to structural changes that make it non-fluorescent. At the level of a single molecule, this process is characterized by an instantaneous interruption of the fluorescence, similar to the light extinction at the breakdown of a light bulb. When observing many molecules at a time, photodestruction results in gradual decrease of fluorescence due to the destruction at each instant of a small fraction of the active molecules.

Fluorescent proteins

One of the major advances in biology and cell imaging in the last twenty years is the development of fluorescent proteins, for which the Nobel Prize in Chemistry was awarded in 2008 to S. Imamura, M. Chalfie and R. Tsien. The first and most famous of these proteins is GFP (Green fluorescent protein), a protein originally discovered and isolated in jellyfish Aequorea victoria [2]. It consists of 238 amino acids arranged in a structure having the shape of a small barrel. From an optical point of view, the GFP absorbs blue light ($\lambda_{exc} \sim 480\,nm$) and emits green fluorescence light ($\lambda_{em} \sim 510\,nm$), which explains its name. The great advantage of GFP is that, by genetic engineering, it is possible to insert its DNA sequence next to that of a protein of interest. Thus, when the protein is produced by the cell, it is directly attached to the GFP and can be detected in a living cell. It is an extremely powerful technique used in a variety of contexts, ranging from bacteria to whole animals. In some organisms, such as bacteria *E. coli*, there are also libraries of mutants in which all genes have been fused to GFP. Since the discovery of GFP, the number of fluorescent proteins has been greatly expanded. Thanks to mutants of GFP and fluorescent proteins from other organisms, there is a wide range of genetically-encoded probes, with emission varying from blue to red [3]. The simultaneous use of several of these fluorescent proteins make it possible to determine the relative position of different proteins and thus to study their interactions. Yet, despite their considerable advantages, fluorescent proteins also have some limitations, including their great sensitivity to photodestruction.

Inorganic nanoparticles

Over the last fifteen years, spectacular progress has been made in the synthesis, characterization, and functionalization of a variety of inorganic nanoparticles. These nano-objects have a size and shape controlled at the nanoscale and their properties depend on the physical nature of the material (metal, semiconductor, magnetic, etc.) which they are composed of. A very promising field of applications is their use as probes in the life sciences. This is in particular the case for semiconductor nanocrystals that have remarkable fluorescence properties [4]. These are particularly intense light emitters, much less sensitive to the photodestruction problems affecting organic fluorescent molecules or proteins. They allow the recording of the movement of biomolecules over large periods, up to several hours. Although these nanoparticles have many advantages in terms of brightness and photostability, they are objects with complex physico-chemical properties and their use in living systems remains often delicate. In particular, the size of the nanocrystals, of the order of 10 to 30 nm, is substantially greater than that ($\sim 1-5\,nm$) of the labeled molecules (see Fig. 2.2). It is therefore crucial to ensure that the nanoparticles do not affect the movement or the biological activity of their target.

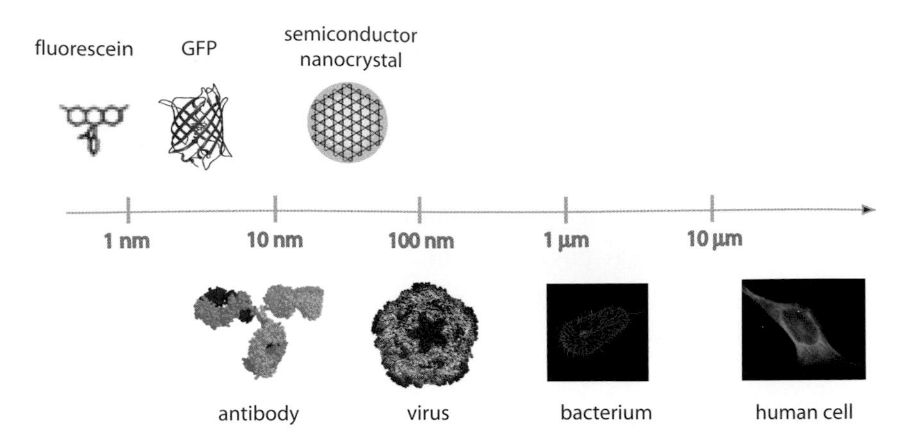

Figure 2.2. Size of the fluorescence probes used in biological imaging in comparison to important elements of living systems.

Box 2.1. Forster energy transfer.

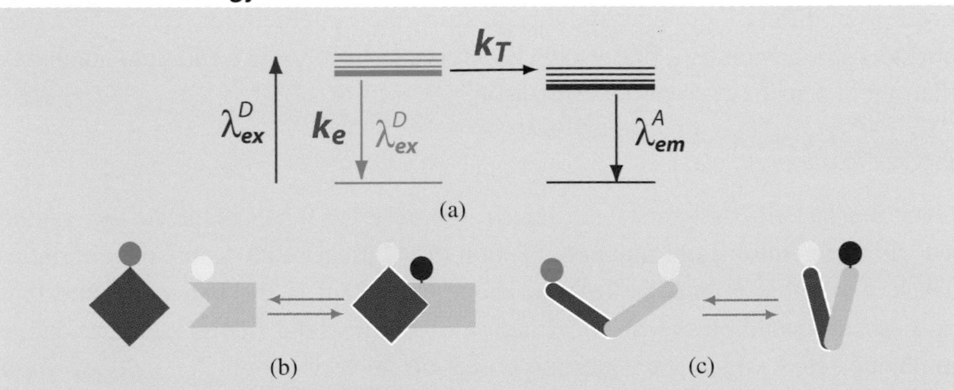

(a) Principle of the energy transfer between a donor and an acceptor molecule.
(b) Application of energy transfer to the study of intermolecular interactions.
(c) Application to the study of intramolecular conformational dynamics.

Labeling molecules with fluorescent probes does not only enable their localization within a biological sample. In some situations, it also gives access to more complex information such as molecular conformations or interactions with specific partners. These measurements are based on a energy transfer mechanism, known as Forster energy transfer. For that purpose, one uses two probes, called donor (D) and acceptor (A) molecules respectively, having the following photophysical properties. The excitation wavelength λ^{D}_{exc} of the donor molecule is such that the acceptor molecule is not excited at this wavelength. In contrast, the emission spectrum (characterized by λ^{D}_{em}) overlaps with the absorption spectrum (characterized by λ^{A}_{exc}) of the acceptor molecule. When the donor molecule is excited, it can relax either by emitting a fluorescent photon or by transferring its energy to the acceptor molecule. In the latter case, the acceptor molecule can then emit a fluorescence photon at the longer wavelength λ_{em}^{A}. The efficiency E of the transfer is determined by the competition between the emission rate k_e of the donor molecule and the transfer rate k_T:

$$E = \frac{k_T}{k_e + k_T}$$

The transfer rate, which reflects the dipole-dipole coupling between the molecules D and A, depends on their relative distance R. More precisely, one shows that k_T scales as $1/R^6$, meaning that:

$$E = \frac{1}{1 + (R/R_0)^6}$$

where R_0 is a parameter which depends on the spectroscopic properties of the probes and is typically on the order of 5 nm. By separately measuring the intensity I_D and I_A at the wavelengths λ_{em}^{D} and λ_{em}^{A}, one can determine the value of $E = I_A/(I_A + I_D)$. Thus, one gets a spectroscopic ruler that connects light intensities to physical distances. Yet, given the spatial dependence of E, the ruler functions correctly only over the range 0.5 R_0–1.5 R_0, that is ~2.5–7.5 nm. Although this range might seem very restricted, it is well adapted to the size of molecular structures. Moreover, in most cases, energy transfer measurements are used not as distance measurements but rather as a binary signal reflecting the proximity, or not, of two molecular partners.

In biology, there are many applications of energy transfer measurements [5]. When the donor and acceptor molecules are bound to different molecular species, it is possible to study their interactions in the context of a living sample. Another application is to couple the molecules (D) and (A) to the same biomolecule. In this case, variations in E are related to conformational changes, possibly induced by the functional activity of the biomolecule.

3 Fluorescence microscopy

3.1 Principle

A microscope is an optical instrument that produces the image of a sample with a high magnification on a sensor [1]. The formation of the image is obtained by means of two optical elements (Figure 2.3): (i) the objective, near of the sample, (ii) the tube lens, placed on the side of the detector. Using a mechanical system, the sample is positioned in a plane, called the focal plane, located at the focal distance f_O of the objective. The objective, which is itself composed of one or more lenses, is used to collect the light from the sample. The light rays coming out of the objective propagate parallel to the optical axis. The role of the tube lens is then to focus these rays and to form an image on a detector located in the image plane, at the distance f_{TL} from the lens.

The performances of an optical microscope are characterized by two important parameters: the magnification and the numerical aperture of the objective. The magnification M represents the ratio between the size of the image on the detector and that of the original sample. G is determined by the ratio f_{TL}/f_O of the focal distances

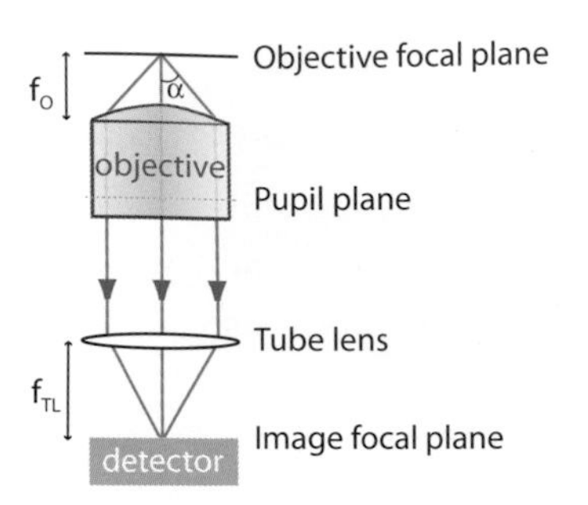

Figure 2.3. Principle of an optical microscope.

of the tube lens and of the objective (Fig. 2.3). The numerical aperture *NA* is defined as $n \sin(\alpha)$ where α is the angle under which the objective collects the light and n is a number (called the index of refraction) that reflects the properties of the medium that separates the objective from the sample or from the glass coverslip on which the sample is deposited. The medium can be just air (in which case $n = 1$) but also water ($n = 1.33$) or oil ($n = 1.5$). As explained below, *NA* also determines the microscope resolution, that is the minimal distance below which two punctual objects in the sample can not be distinguished in the image. In most microscopes, the tube lens is fixed and only the objectives can be changed. It is thus common to indicate directly on the objective its *NA* and the magnification *M* (although the latter implicitly depends on the tube lens' properties). In commercial microscopes, the magnification can be as high as 200 and the *NA* can reach 1.49 for oil-immersion objectives.

For fluorescence imaging, a light source, usually a lamp or a laser, is used to excite the sample whose fluorescence is collected to form an image. In practice, an important benefit of fluorescence imaging lies in the possibility to spectrally discriminate between the excitation and emission light. By means of optical filters placed between the sample and the detector, one selects the emission fluorescence and rejects the excitation light. Just as it is simpler to observe stars at night than during daytime, the sensitivity of fluorescence measurements is largely increased by the fact that detection occurs on a dark background.

A fluorescence microscope can be schematically separated into two parts: (i) an excitation path by which the light from the source is sent onto the sample, (ii) an emission path by which fluorescence photons are sent from the sample onto the detector. These two paths are combined thanks to a dichroic mirror placed between the objective and the tube lens, which reflects the excitation light and transmits the fluorescence emission. Based on this general principle, there exists two broad classes of fluorescence microscopes: wide-field epifluorescence microscopes and confocal scanning microscopes.

3.2 Wide-field epifluorescence microscopy

The simplest imaging method is the wide-field epifluorescence microscopy. It consists in focusing the excitation light in a particular plane, called the back focal plane of the objective (Fig. 2.4a1), so that the sample is illuminated by a parallel beam with a large spatial extent. In this configuration, the entire sample (for example, a cultured cell on a glass slide) is excited and all the fluorescent probes emit simultaneously light. The image of the sample is recorded on an extended detector, for example a digital camera.

A central issue in microscopy is to determine the spatial resolution of an optical instrument. To do so, the easiest way is to first consider the case where the sample probe contains a point object, or more realistically, an object having a size much smaller than wavelength λ_{em}. According to the laws of wave optics, the image of this source is no longer punctual, but rather a spot whose lateral extension is about $d = \lambda_{em}/2NA$ multiplied by the magnification M. As a result, two point sources in the sample can be distinguished in the image only when their relative distance is greater than d. In other words, when looking at a fluorescent sample, which can be considered as composed of a large number of point sources, it is not possible to access information on details at a scale smaller than the resolution d. This law bears the name of Abbe who first derived it in the 19th century. The numerical aperture also determined the depth of focus $p = \lambda_{em}/4n(1 - \sqrt{1 - NA^2/n^2})$ that is the sample thickness over which the microscope produces a focused image. In the common limit where $NA \ll n$, p is equal to $\lambda_{em}n/2NA^2$. When a point source is out of focus, that is located at a distance of the objective focal plane greater than p, it appears as a defocused light spot with a spatial extension larger than d (Fig. 2.4b).

Although the epifluorescence microscope is a very widely used instrument, it suffers from an important limitation when samples thicker than the depth of field p are studied. Since the entire sample is illuminated by the excitation light, the detector records the light coming not only from the part of the sample within the depth of focus but also from off-focus parts. The latter add a large amount of background light that degrades the sensitivity of the detection and the quality of the image. There are several solutions to overcome this difficulty and reject off-focus photons. A first approach consists in using an excitation pattern localized in space. To do so, one takes advantage of the difference between the index of refraction of the glass coverslip ($n_{glass} = 1.5$) on which the sample is deposited and that of the aqueous buffer ($n_{water} = 1.33$) in which it is incubated. Due to the Snell–Descartes law of refraction at the glass-water interface, a light ray with an incident angle θ larger than the critical angle $\theta_c = \arcsin(n_{water}/n_{glass})$ is not transmitted from the coverslip to water

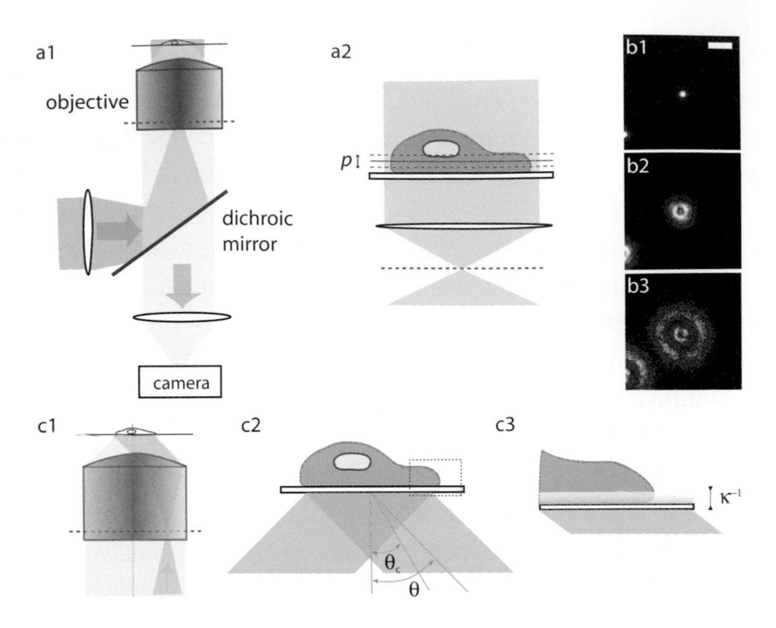

Figure 2.4. Fluorescence microscopy. (a1) Schematic of an epifluorescence microscope. (a2) The entire sample is illuminated with an extended beam. The plain line corresponds to the objective focal plane and the dotted lines indicate the depth of field p of the microscope. (b) Image of a point source placed at a distance L from the focal plane and acquired using an objective with magnification $M = 100$x and $NA = 1.4 : L = 0$ (b1), $L = 500$ nm (b2), $L = 1\,\mu$m (b3). The scalebar corresponds to 1 μm. (c1) Schematic of an evanescent-wave microscope. The laser focal spot is laterally shifted in the back focal plane of the objective (dotted line) in order to create an oblique illumination. (c2) When the incidence angle θ is larger than the critical angle θ_c, the excitation beam is reflected at the glass-water interface. (c3) At the interface, there is an evanescent field with a decay length κ^{-1}.

but is entirely reflected. Yet, at the interface, there is an evanescent electric field, the intensity of which decays as $\exp(-\kappa z)$ where $\kappa = (4\pi/\lambda)\sqrt{n_{\text{glass}}^2 \sin^2(\theta) - n_{\text{water}}^2}$. Therefore, this field can be used to selectively excite fluorescent molecules located within a distance on the order of κ^{-1} from the glass coverslip.

In an epifluorescence microscope, an evanescent wave excitation can be achieved by focusing a laser beam in the objective back focal plane at a position shifted from the optical axis (Fig. 2.4c). This lateral shift results in an oblique illumination in the objective focal plane. If the NA of the objective is such that the angle α is larger θ_c, the illumination angle can become larger than the critical angle by shifting the laser focal position sufficiently far from the center. The decay length κ^{-1} of the evanescent wave is usually on the order of 100 nm. Consequently, evanescent wave microscopy is limited in cell biology to the investigation of processes occuring at or very close to the plasma membrane, such as the traffic of membrane proteins or the release of neurotransmitters in nerve cells.

3.3 Confocal scanning microscopy

Confocal microscopy, invented by Marvin Minsky in 1957, is a general technique and effective in removing the out-of-focus background signal in fluorescence images. In a confocal microscope, the sample is excited not with an extended illumination but by focusing a laser beam in the objective focal plane (Fig. 2.5). The trick is to add in the image plane an opaque mask containing a small filtering pinhole with a diameter $\sim Md$. Using an additional lens, the light passed through the pinhole is detected on a photodetector. Thus, if a light source is exactly in the focal plane of the microscope, its image will pass through the hole and the photons arrive at the detector. In contrast, if it is at a distance comparable to or greater than p, its image is a wide spot which is almost entirely blocked by the mask (Fig. 2.5a3 and a4). Thus, with the mask one selects the photons coming from a zone of thickness $\sim p$ around the focal point and rejects all the off-focus light. In other words, the filtering hole defines a volume around the laser focal point and only molecules in this volume will contribute to the signal. To form an image of the sample, the focused beam is moved laterally and at each position the fluorescence signal is recorded on the detector. Thus, in contrast to the epifluorescence microscope for which there is a parallel detection of all the pixels of the camera, the confocal microscope is a scanning system for which the detection is sequential. Importantly, the resolution of a confocal microscope is primarily determined by the extension of the focused beam (and not the size of the filtering hole) and is at best also in the order of d.

Confocal microscopes enable other types of useful measurements in which the beam is not scanned. Rather, the laser spot is fixed and one records the time course of the fluorescence signal generated by the molecules coming across the focal volume. At the microscopic level, biomolecules primarily move by Brownian diffusion.

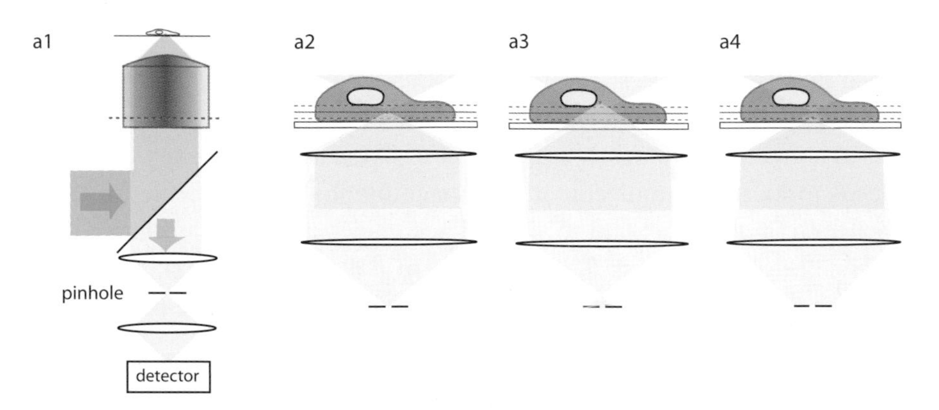

Figure 2.5. (a1) Schematic of a confocal microscope. (a2–a4) According to the source position with respect to the focal plane and the depth of field, the fluorescence signal is blocked or not by the pinhole.

During diffusive processes, the motion is random and, as a result, the number of molecules in the focal volume varies. These variations lead to temporal fluctuations of the fluorescence intensity from which useful information can be deduced. To this end, one computes the normalized autocorrelation function:

$$g(\tau) = \frac{\langle I(t)I(t+\tau)\rangle_t}{\langle I(t)^2\rangle_t}$$

($\langle . \rangle_t$ denotes time averaging). When the motion of the fluorescent molecules is solely governed by diffusion (with a diffusion coefficient D), the autocorrelation function is equal to:

$$g(\tau) = 1 + \frac{1}{N}\frac{1}{1 + \tau/\tau_D}$$

where N is the average number of molecules in the focal volume and $\tau_D \propto 1/D$. Thus, fluorescence correlation measurements give access to the concentration of molecular species as well as to their transport parameters.

4 Single molecule imaging

4.1 Imaging and localization of an individual probe

Over the last fifteen years, a significant effort in optical microscopy has been conducted to increase the detection sensitivity in fluorescence measurements. To this end, it is necessary to optimize the collection of the fluorescence signal while minimizing the noise (see box on the concept of signal to noise ratio). This effort has been largely successful since it is now possible to detect the signal from a single probe attached to a biological molecule of interest in a living cell. This has opened fascinating perspectives for fluorescence imaging and its applications in biology and biotechnology.

Since fluorescent probes (organic molecules, fluorescent proteins or inorganic nanoparticles) are all small compared to λ_{em}, they behave as point sources. As discussed above, their image is a light spot of extension d. In many experiments, one needs to locate the molecule in the image by pointing the center of the spot. In practice, it is possible to do so with an accuracy σ much better than d. In fact, the larger the signal (that is the number N of collected photons), the better the localization accuracy (see Fig. 2.6). In the so-called photon noise limit (see box), σ is on the order of $d/\sqrt{N}$ [6].

If, for instance, one detects in an image 1000 visible photons ($\lambda_{em} = 500$ nm) per molecule using an objective with $NA = 1.4$, the localization accuracy is on the order

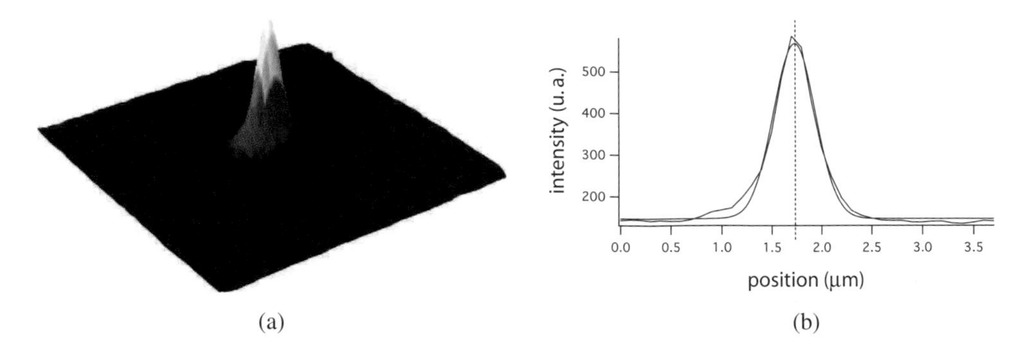

Figure 2.6. (a) Three-dimensional representation of the intensity profile in the image of point source. (b) Adjustment of the intensity profile (in blue) by a Gaussian curve in order to determine the source position (dotted line) with an accuracy $\sim d/\sqrt{N}$ inferior to the width of the profile (on the order of d).

of 6 nm, close to the molecular scale. Yet, we emphasize that localizing a molecule with an accuracy σ less than d does not mean that we increased the resolution, that is, we gained the capacity to distinguish two molecules having a relative distance smaller than d. To accurately localize a single molecule, it is essential to know that the signal only comes from a single molecule!

Box 2.2. Signal to noise ratio.

The sensitivity of an optical measurement is determined by the signal to noise ratio. The signal S results from the sum of the signal S_f of the fluorescent probe and of the background signal S_b due to the stray light reaching the detector or to the electronic circuit that measures the light intensity. To evaluate the noise, it is necessary to determine the fluctuations of the signal by computing the variance σ_S of S. Since the fluctuations of the fluorescence signal and of the background are independant, σ_S is given by the relation $\sigma_S^2 = \sigma_{S_f}^2 + \sigma_{S_b}^2$ where σ_{S_f} and σ_{S_b} correspond to the variance of the fluorescence signal and of the background, respectively.

The average fluorescence signal per molecule is equal to $\bar{S}_f = \Gamma T$ where Γ is the fluorescence emission rate and T is the duration of the acquisition. Yet, there is a noise associated to this signal. Even if the rate Γ is fixed, the measured signal fluctuates at each time interval T. These fluctuations, called photon noise, are such that $\sigma_{S_f} = \sqrt{\Gamma T}$. The average value of S_b and its variance σ_{S_b} are usually determined by measuring the signal in the absence of fluorescent probes. Thus, we deduce the expression of the signal to noise ratio:

$$S/B = \frac{\bar{S}_f}{\sqrt{\sigma_{S_f}^2 + \sigma_{S_b}^2}} = \frac{\Gamma T}{\sqrt{\Gamma T + \sigma_{S_b}^2}}$$

In the case where ΓT is much larger than $\sigma_{S_b}^2$, one reaches the shot noise limit $S/B \simeq \sqrt{\Gamma T}$. It is an intrinsic limit in optical measurements, which does not depend on the detection conditions or the technical features of the detector but is due to the corpuscular aspect of light.

4.2 Super-resolution photoactivation microscopy

In biological imaging, the resolution $d = 2\lambda_{em}/NA$ of microscopes is at best on the order of 200 nm. This length scale should be compared to the size of a cell (1 to 100 μm), of an organelle ($\sim$0.1 to 1 μm) or to many macromolecular assemblies ($\sim$10 to 100 nm). In other words, optical microscopy is a powerful tool to access information at the cellular scale but it is less appropriate to probe molecular structures. To reach the molecular and subcellular scale, it is thus necessary to break the Abbe diffraction limit. In the last years, several teams have shown how this can be achieved by using clever excitation and detection schemes [7]. Here, we will limit our discussion to photoactivation localization microscopy, a technique invented in 2006 by the groups of E. Betzig, X. Zhuang and S. Hess, and based on the detection of individual molecules.

The principle of photoactivation localization microscopy relies on the use of molecules that can be activated or deactivated, i.e. molecules which can be transfered from a non-fluorescent (dark) states to a fluorescent (bright) state. By activating a sufficiently small subset of the molecules within the sample, it is possible to reach a regime where the average distance between activated probes is larger than d. Thus, the fluorescence image is composed of isolated fluorescent spots coming from individual molecules. Each spot can be analyzed in order to localize the corresponding emission source with an accuracy σ. To image the full sample, it is then necessary to sequentially activate a few molecules, localize them and deactivate them. By repeating this sequence of events a large number of times (ideally until all the fluorescent molecules in the field of view have been detected), one reconstructs an image whose resolution is not given by the Abbe limit anymore but by the localization accuracy. In practice, σ is varied between 10 and 50 nm. This method, called PALM (Photo-Activation Localization Microscopy) or STORM (Stochastic Optical Reconstruction Microscopy) in the literature, is somehow analogous to the pointillist technique developed by artists in the late 19th century in which the image in the painting results from the accumulation of dots of paints on the canvas (corresponding in super-resolution microscopy to the position of individual fluorescent molecules).

A key element in photoactivation microscopy is to have a mechanism for controling the activation (turn-on) and deactivation (turn-off) of fluorescent probes. This can be nowadays achieved for many different types of fluorescent markers. An important case is that of photoconvertible proteins. These proteins, which emit in their non-activated form at a certain wavelength (usually in the green region of the visible spectrum), shift their emission wavelength towards the red after activation by violet light ($\lambda \sim 400$ nm). When imaging in the green, one gets a conventional

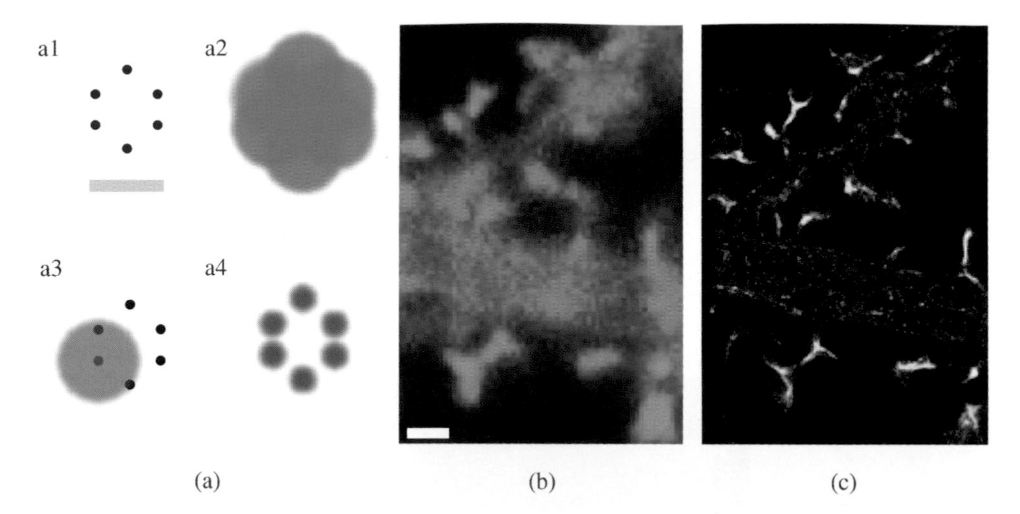

Figure 2.7. (a) Principle of super-resolution photoactivation localization microscopy. (a1) Position of the probes in the sample. The scale bar indicates the resolution d of the microscope. (a2) Simultaneous image of all the probes. (a3) Only one of the molecules is activated and its localization is determined by finding the center of the fluorescence spot (in red). (a4) Reconstructed image after sequential activation of all the molecules. The size of the spots designates the localization accuracy. (b) Conventional fluorescence image of the actin cytoskeleton in cultured neurons. The image resolution is $\sim$200 nm. (c) Image by photoactivation localization microscopy. The resolution is on the order of 40 nm. Scale bar: 1 μm (adapted from [8]).

microscopy image since all the molecules emit simultaneously. After a short flash of violet light, a few molecules are photoconverted and can be detected as individual molecules by collecting fluorescence in the red. Subsequently, the activated molecules are switched off by photodestruction. The main advantage of these photoconvertible probes is that, similarly to GFP, they can be fused to proteins of interest and enable the investigation of the distribution of specific proteins in living cells with nanoscale resolution. Figure 2.7b and c illustrates in the case of the actin cytoskeleton in neuronal cells the gain in resolution made possible by PALM/STORM microscopy.

4.3 Single molecule tracking

Another important application of single molecule imaging is the tracking of individual proteins in cells. In this case, the fluorescence signal of an individual molecule is recorded in a sequence of images. In each image, the localization of the molecule is determined with sub-diffraction accuracy. Analysis of the trajectory, formed by the consecutive positions of the molecules, gives access to important parameters

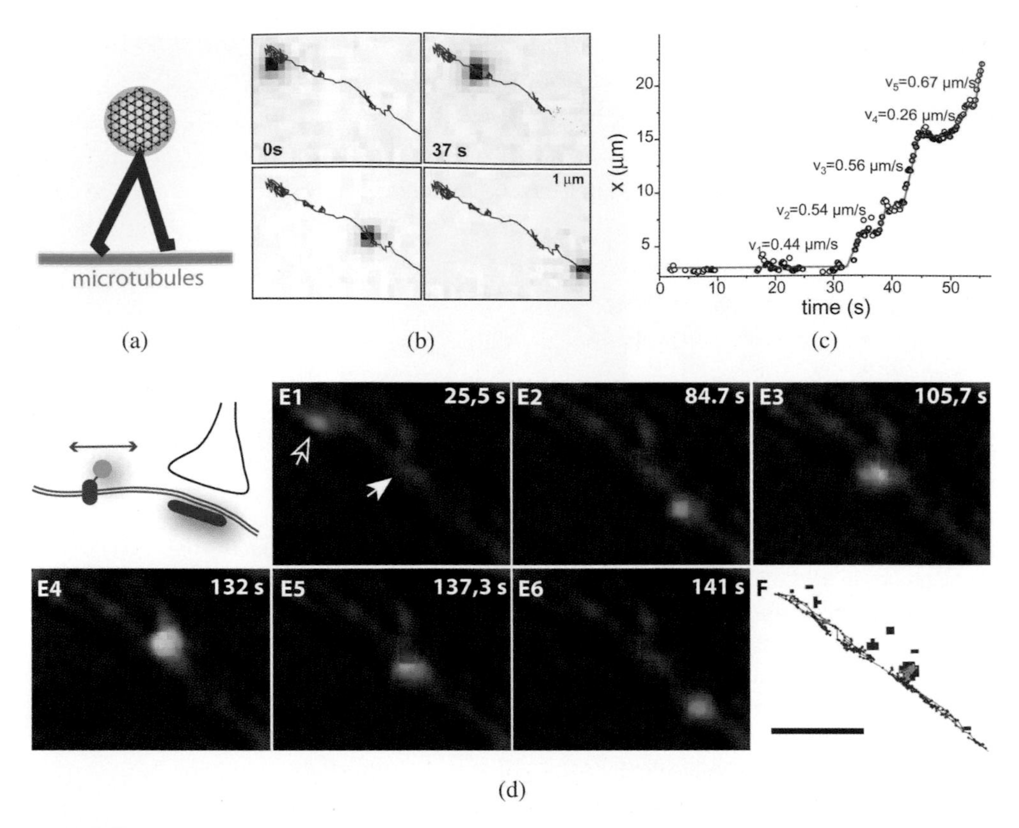

Figure 2.8. Tracking of individual molecules in living cells. (a) Example of Kinesin, a molecular motor moving along microtubules. (b) Sequence of images showing the directed motion of a motor coupled to a fluorescent nanoparticle inside a living cell. (c) The analysis of the motion indicates that the motor alternates between phases of directed motion and pauses (adapted from [9]). (d) Example of Glycine receptors (blue) coupled to a fluorescent probe (green). The receptor diffuses in the neuronal membrane and reaches synaptic sites where it is stabilized by scaffolding proteins (red). (e1–6) Sequence of images showing the movement of a single glycine receptor (green) with respect to a synaptic site (red). (f) Reconstruction of the receptor trajectory (blue). The green points indicate the parts of the trajectories when the receptor is trapped at the synapse (adapted from [11]).

underlying the spatial dynamics, such as the diffusion coefficient of the transport velocity.

We illustrate here the interest of single molecule tracking tools in cell biology with two examples. In the first one, one is interested in the movement of kinesin, a molecular motor that moves along microtubules and plays an important role in intracellular transport and cell division (Fig. 2.8a–c). The properties of this motor have been extensively studied in *in vitro* assays, using in particular optical tweezers measurements (see Chapter 3). These experiments have given access to the speed of the motor as well as to the forces (on the order of 5 pN) that it can generate during transport of a cargo. However, the cytosol is an environment very different from the one encountered *in vitro*. For instance, it is crowded with molecules that can bind to microtubules and act as obstacles for the kinesin motion. By labelling kinesins with labels (such as a semiconductor nanoparticle) and introducing them inside cells, one can record their movement at the single molecule level [9] and directly measure their velocity (on the order of 0.5 μms^{-1}) and processivity (on the order of 1 s), that is the duration of their directed movement along microtubules before their unbinding. Surprisingly, the results are very similar to those found *in vitro*, suggesting that molecular obstacles are certainly not very stable and that they rapidly dissociate from microtubules. A second example is related to the membrane dynamics of glycine receptors, a neurotransmitter receptor involved in the inhibition of the nervous signal (Figs. 2.8d–f). The localization of the receptors at synapses, and its regulation, is an important question for our understanding of the plasticity of the nervous system. Using single molecule imaging, it has been observed that the receptors diffuse rapidly in the membrane. Moreover, they are much less stable than expected at synapses, where they only transiently reside.

5 Conclusion and perspectives

Optical microscopes enable the investigation of biological systems with an ever increasing sensitivity. Coupled with fluorescence labeling tools, imaging techniques give access to the localization of molecular actors involved in a multitude of biological processes. In recent years, a special effort has been made to measure the properties of living systems. These measurements that make a link between the functional role and spatial dynamics of molecules, organelles and cells, continue to raise new questions about the organization of living matter. In this chapter, we focused on the description of imaging tools for cell biology. However, many important questions in biology are related to the properties of multicellular systems, such as tissues, brain slices or whole organisms. From an imaging perspective, these are thick samples wherein the propagation of light is complex. They pose many challenges to which microscopists respond with more developed advanced techniques, such as nonlinear optical imaging (see Chapter 6) or adaptive optics.

Beyond fundamental research, fluorescence microscopes are also of great use for medical diagnosis. For instance, when implemented in an endoscope, they can measure the properties of tissue directly from a patient and will probably play a role as guiding tools for surgeons. Microscopes are not only tools for the observation of biological systems, but also for their mechanical or chemical manipulation. One can for example apply mechanical forces with optical tweezers (see Chapter 3) or focus powerful lasers for the nano-dissection of tissue. Using photoresponsive chemical probes, it is possible to trigger biochemical reactions, by changing for instance the local concentration of ions or by forcing the interaction between molecular partners. In all these cases, microscopy plays a dual role, serving both to generate well-controlled perturbations and to measure their effects in biological samples.

Biological imaging experiments are becoming ever more complex, not only from an instrumental point of view but also for the image analysis. As illustrated in the case of photoactivation localization microscopy, acquiring an image depends not only on recording an optical signal but also on the digital processing of the data. In addition, dynamic measurements in microscopy generate increasingly important quantities of data. In developmental biology, one would like to record the full development of an embryo (Zebrafish or Drosophila, in particular) for periods of up to 24 hours or more. To achieve the spatial ($\sim$1 μm) and temporal ($\sim$1 mn) resolution necessary to follow in the embryo the flow of cells and their division cycles, one must acquire several hundreds of thousands of images, corresponding to Terabytes of data. [12] One easily conceives that handling such an amount of data and retrieving relevant information necessitates specific skills. In conclusion, biological imaging is a multidisciplinary field of research which require the joint effort of biologists, chemists, physicists and computer scientists. In the future, it is by combining all these skills that we will continue to shed some light on the complexity of living matter.

Bibliography

[1] An introduction to microscopy techniques is available on the web site: http://micro.magnet.fsu.edu/primer/index.html. For a detailed description: J. Mertz, *Introduction to Optical Microscopy*, Roberts and Editors (2009).

[2] A presentation of the discovery and applications of GFP is avaiable on: http://nobelprize.org/

[3] Tsien RY, *FEBS Lett.* (2005) **579**, 927–932.

[4] Michalet X, Pinaud FF, Bentolila LA, Tsay JM, Doose S, Li JJ, Sundaresan G, Wu AM, Gambhir SS, Weiss S. *Science* (2005) **307**, 538–44.

[5] Selvin PR. *Nat. Struct. Biol.* (2000) **7**, 730–4.

[6] Thompson RE, Larson DR, Webb WW. *Biophys J.* (2002) **82**, 2775–83.

[7] Huang B, Babcock H, Zhuang X. *Cell* (2010) **143**, 1047–58.

[8] Izeddin I, Specht CG, Lelek M, Darzacq X, Triller A, Zimmer C, Dahan M, *PLoS One* (2011) **6**, e15611.

[9] Courty S, Luccardini C, Bellaiche Y, Cappello G, Dahan M. *Nano Lett.* (2006) **6**, 1491–5.

[10] Dahan M, Lévi S, Luccardini C, Rostaing P, Riveau B, Triller A. *Science* (2003) **302**, 442–5.

[11] Ehrensperger MV, Hanus C, Vannier C, Triller A, Dahan M. *Biophys. J.* (2007) **92**, 3706–18.

[12] Keller PJ, Schmidt AD, Santella A, Khairy K, Bao Z, Wittbrodt J, Stelzer EH. *Nat. Methods* (2010) **7**, 637–42.

3

Mechanical studies on single molecules: general considerations

David Bensimon, *CNRS researcher, Laboratoire de Physique Statistique de l'ENS, Paris.*
Vincent Croquette, *CNRS researcher, Laboratoire de Physique Statistique de l'ENS, Paris.*

1 Elements of molecular biology

The cell is a microscopic biochemical factory of great complexity. Enclosed by a proteo-lipidic membrane, the cell contains a large number of protein complexes (ribosome, centriole, etc.) and organelles (mitochondria, chloroplasts, Golgi apparatus, etc.) whose functions are coordinated to ensure cell survival in various environments (Fig. 3.1). Its core, the cell nucleus, contains the instructions required to fulfill its functions and ensure its identical reproduction. The nucleus consists essentially of one or more DNA molecules compacted by their winding around protein complexes (histones), positioned along DNA as pearls on a necklace. The DNA molecule itself consists of a pair of complementary hetero-polymers made of four different types of monomers (or bases): Adenine (A), Guanine (G), Cytosine (C) and Thymine (T) whose positions along the polymer determine its sequence (e.g. … AGGATTCGGAAT …) (see Fig. 3.3). The two strands of the DNA molecule form a double helix: the bases A (G) on one strand are paired with the bases T (C) on the other. DNA is the memory of the cell, its reproduction requires replicating the molecule (copying the memory), a process facilitated by the very

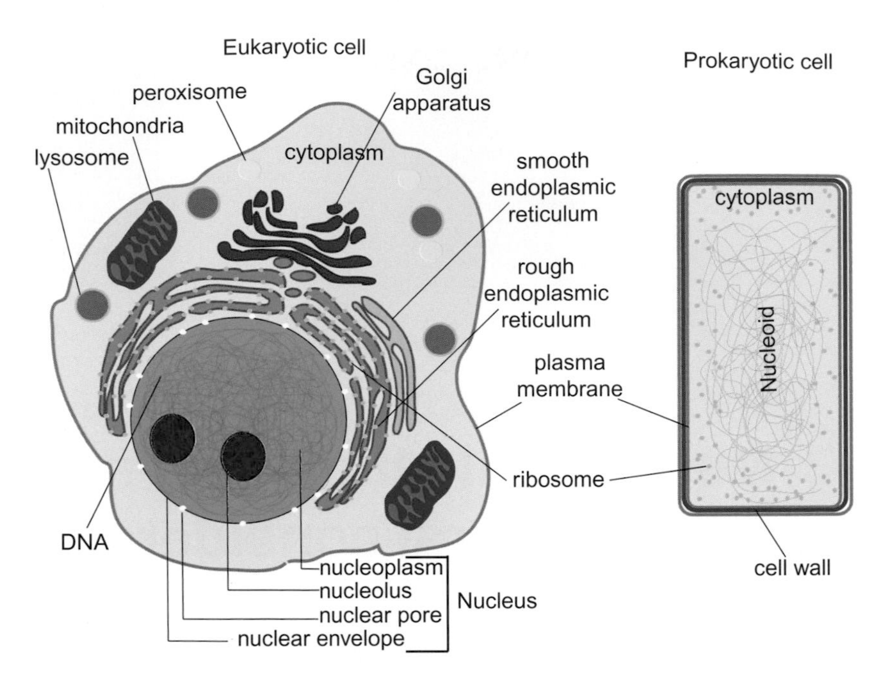

Figure 3.1. Schematic diagram of an animal cell. Note the nucleus and the existence of a large number of protein units specialized in certain tasks: the ribosome for peptide synthesis, the mitochondria for ATP synthesis, microtubules for intracellular transport, the Golgi apparatus for secretion, etc.

structure of DNA: a strand serving as template for its complementary copy. A small portion of the DNA sequence codes for the structure of the various proteins involved in the operation of the cell. However most of the sequence is used to regulate the expression of these coding sequences and therefore control the protein content and the cellular response to a particular environment. The central role of DNA in the cell implies that it interacts with a large number of proteins. Some of these proteins (such as histones) are structural, others are molecular motors that open the molecule (helicases), copy it (DNA polymerases), translate it into messenger mRNA (RNA polymerases), unknot it (topoisomerases), repair it (such as the MutS-MutL complexes), translocate it (such as the protein FtsK), etc. The development of techniques for manipulating single DNA molecules (described below) has allowed the study of these interactions at the single molecule level: the study of a single protein interacting with a single DNA molecule. These investigations have led to a better and more detailed understanding of the processes that involve DNA.

If DNA is the genetic memory of the cell and the proteins its specialized "workers", RNA has emerged in recent years as its "handyman" molecule. Its first identified role was that of a transient copy. According to the "central dogma" of molecular biology (see Fig. 3.2) the information coded in DNA flows in one direction: it is

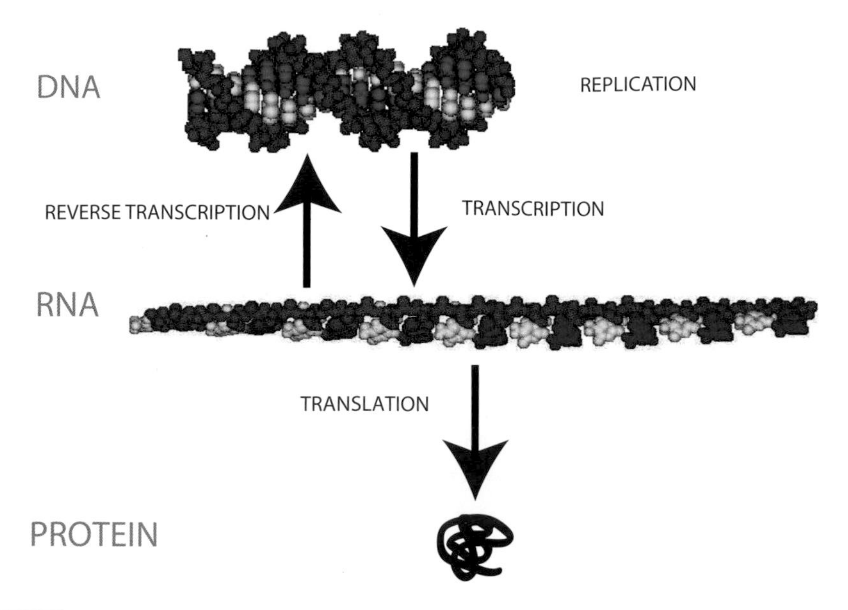

Figure 3.2. The central dogma of molecular biology: genetic information flows from DNA to mRNA, which can fold into secondary structures and on to proteins. However, some exceptions exist (reverse transcriptases can encode DNA from RNA). (Pictures courtesy of R. Lavery.)

first copied into mRNA, which is then translated into a peptide sequence which folding in 3D defines the structure of the protein encoded in the original DNA sequence (or gene) (each triplet of bases defining one of 20 amino acids composing the protein). It then appeared that the translation process required some specific RNAs, known as transfer or tRNA which are used as adapters between the sequence of the mRNA, that they recognize specifically (they pair as in DNA with successive triplet of bases (codons) and the corresponding amino-acid (which is linked at one end of the tRNA). This translation process is coordinated by a huge nucleo-protein complex: the ribosome that interacts with both mRNA and tRNA. This complex consists of two subunits, each comprising several tens of proteins that form a complex with yet other specific RNA molecules, ribosomal RNA or rRNA. It has emerged in recent years that the catalytic function of the ribosomes (the formation of a peptide bond between nearby amino-acids) was in fact mediated by rRNA, while the protein complex itself was playing an essentially structural role. In fact many RNA molecules with catalytic activity, so called ribozymes, have been discovered.

Beyond these functional roles of RNA in transcription and translation, it has recently appeared that small RNA molecules also play a role in regulating genetic expression and translation. These multiple roles of RNA reinforce the hypothesis of the RNA world as an origin for Life. RNAs would have appeared first,

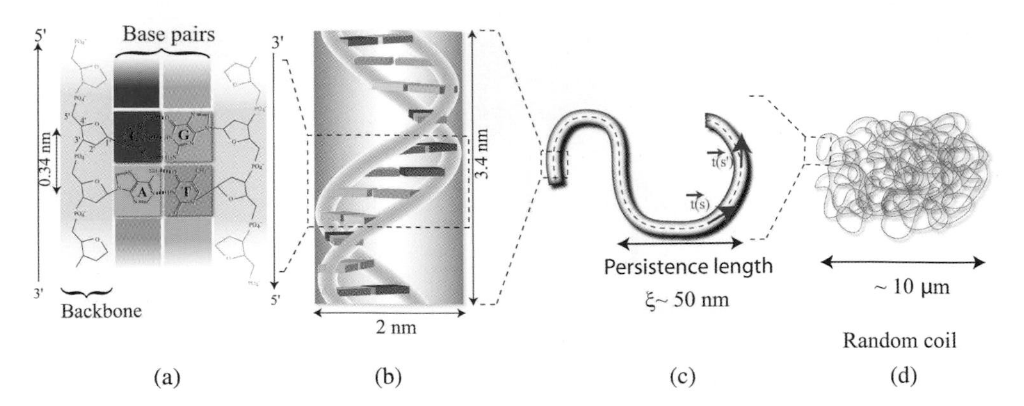

Figure 3.3. DNA structure. (a) Each strand is composed of a sugar-phosphate backbone. Chemical groups (Adenine (A), Guanine (G), Cytosine (C) and Thymine (T)) are attached to the sugar at position 1'. For the standard form of DNA ("B-DNA"), the distance between base pairs is 0.34 nm. The two strands are anti-parallel: the orientation can be defined using the points of attachment of the bases to the sugar, (b) secondary structure of DNA: each strand is wound around each other to form a right handed helix. The pitch of the helix is about 10.5 base-pairs, which implies that the twist angle between bases is about 36°. At larger scale, DNA behaves as a semi-flexible chain which orientation is uncorrelated over a distance of $\xi = 50\,\text{nm}$, called the persistence length. (d) Long DNA molecules (length $L \gg \xi$) behave as random coils with an average size R that increases as the square-root of the length: $R = \sqrt{2\xi L} \sim 10\mu\text{m}$ for the chromosome of *E. coli* which has 4.6 million base pairs (length $L \sim 1$ mm). (Figure inspired by [6].)

functioning both as a memory and a catalyzer for their own replication. DNA would have appeared later, a hypothesis supported by the discovery of enzymes (reverse transcriptases) that synthesize a complementary DNA from an RNA substrate. Finally, proteins with varied and specific functions would have appeared catalyzed by ribozymes or some primitive form of the ribosome.

Apart from the nucleus and its DNA, the cell contains many organelles which functions can be studied at the single molecule level. The targeting of certain proteins (such as synaptic receptors that have to reach the synapse several tens of cm away from the cell) often does not take place by passive diffusion but in micro-vesicles transported by molecular motors (kinesins, dyneins, etc.) along cellular rails: microtubules. In addition, the cell is not a bag of enzymes (and organelles). It is a structured viscoelastic medium underpinned by actin fibers forming an active cytoskeleton involved in the movement of the cell. The study of these motors and the various cellular networks of fibers (actin, microtubules, etc.) has largely benefited from the development of the micro-manipulation techniques described next page.

2 Advantages and drawbacks of single molecule studies

Ever since its origins, biology has been enriched by the contribution of new techniques enabling detailed observation of living systems. Thus the optical microscopy techniques developed by Leeuwenhoek and Hooke in the 17th century, laid the groundwork for cellular biology. While these techniques have continuously improved, the classical microscope based on linear optics cannot resolve features smaller than 200 nm. Recent microscopy approaches using the nonlinear optical response of fluorophores, have managed to essentially abolish the so-called Abbe-limit, reaching a resolution of 20 to 30 nm.

The contribution of electron microscopy, developed in the mid 20th century, allowed for the observation of objects with even higher resolution (up to fraction of a nanometer for a transmission electron microscope (TEM)). However, while optical microscopy allows to obtain images of live samples in physiological conditions (in aqueous solutions), electron microscopy requires conditions far removed from physiological ones, such as freezing samples to very low temperatures (so-called CryoEM techniques).

Other techniques are characterized by even higher resolution, for example X-ray diffraction was used to study the structure of DNA and deduce its famous double-helical structure. Despite the high resolution of this technique (on the order of 0.1 nm), it requires the use of crystallized samples, which are static and often obtained in conditions far from physiological ones.

NMR spectroscopy, which appeared in the middle of the 20th century, has a very high resolution (it can resolve molecular structure) and it can study molecules in a physiological aqueous solution, but it is a rather slow and insensitive technique that necessitates a huge amount of material (e.g. proteins) to get a high enough signal to noise ratio.

Although these imaging techniques provide considerable information, most experiments aiming to understand the mechanisms of interaction between proteins and DNA were conducted in test tubes, on large ensembles of molecules.

In the past twenty years, technical progress in optical microscopy methods and the advent of single molecule micromanipulation methods have fulfilled an old dream: to both visualize and manipulate a single molecule, in real time and in physiological conditions. In what follows, we will first explain what type of information such an approach can yield. We will the review the orders of magnitude of forces acting at the molecular level and describe some of the techniques used to manipulate biomolecules and DNA in particular.

2.1 Single molecule

Consider a typical experiment in biochemistry. It involves a test tube containing a protein solution at a typical concentration of 100 nM in a volume of 100 μL. A quick calculation shows that this test tube contains approximately one billion proteins. Therefore, any measurement performed on the contents of the tube accesses only the average properties of this population.

This bulk measurement is insensitive to two types of variability present in the population, so-called static and dynamic disorder. The static disorder reflects the heterogeneity of a population of molecules. The proteins in a test tube, even if they are identical (i.e. have the same amino-acid sequence) may have different properties. The proteins could for example exist in two different states (two different folding conformations of their amino-acid sequence), each state with specific biochemical properties. If one measures some property of the proteins, the result will be an average value of that property in the two states, which may be very different from the value in each state (Fig. 3.4). In particular, the presence of inactive proteins in a sample — which are often indistinguishable from active ones — inevitably leads to an underestimate of the activity of a protein.

Dynamic disorder stands for the time variation of the properties of a given molecule as it switches between various states. For example, the catalytic rate of an enzyme may vary over time, as it switches between different catalytic conformations. However, in a test tube, the molecules are not synchronized, and the measurement of a given property will again be an average over its variation (Fig. 3.4(b)).

In addition to overcoming the adverse effects of ensemble averaging, single molecule visualization and manipulation techniques often provide more data and greater details than bulk biochemical assays. Thus, in the case of helicases, the enzymes unwinding the two strands of DNA, bulk measurements sense only two states: fully unwound (separated) DNA or duplex strands. Various single molecule techniques can however monitor in real time the unwinding of a DNA substrate by a single helicase, opening a vista on the intermediate states between the initial duplex DNA and the final state when the two strands are completely separated. The data obtained on the catalytic cycle of enzymes is thus much more detailed. In particular, the step-size of a molecular motor (the distance translocated on a substrate such as DNA in a single catalytic cycle) and its processivity (the total distance travelled before falling off) can be deduced from single molecule measurements. These parameters are very difficult to get from bulk studies (even though stop flow studies by synchronizing the molecule partially address that issue).

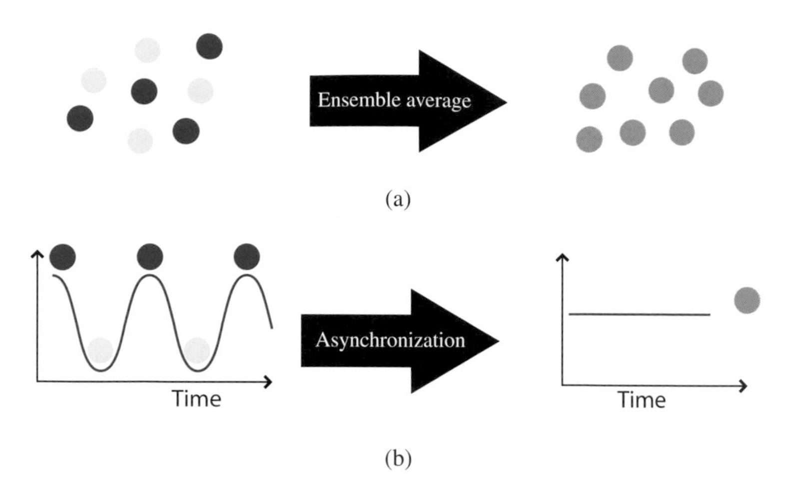

Figure 3.4. (a) Average static disorder. An experiment in a test tube, on a large population of molecules, yields the average of a property, not its distribution. Here two populations coexist, each with specific biochemical properties (symbolized by different colors). As a result, the average of the two populations does not offer a correct description of their properties at the microscopic level. (b) Average dynamic disorder. A molecule can occupy two distinct states over time, for example two different conformations. Each state is characterized by specific biochemical properties (symbolized by different colors). The result of the measurement of the quantity of interest in a desynchronized population is the average property in the two states, which again does not reflect the distinct properties of each state, or the dynamics of state switching.

Finally the single molecule approach allows for observing effects that would be impossible to investigate otherwise, as for example the rotation of the sub-domain γ of the F1-ATpase, which is driven by the hydrolysis of ATP to ADP but can also be shown to generate ATP from ADP when rotated by an external force (see Chapter 4).

Even though studies on single molecules provide a detailed description of the molecular behavior within a population, correlating these results with bulk measurements is usually a long and tedious process. As the size of the studied population is often well below that of bulk investigations, the statistical error on the average is much greater. If the population is relatively homogeneous, it is therefore generally preferable to use bulk approaches over single molecule ones to get more precise estimates.

3 Order of magnitude of the relevant parameters at the single molecule level

It is important, prior to a description of the various single molecule manipulation techniques, to recall the order of magnitude of the effects that intervene at the molecular level.

3.1 Energies

The relevant order of magnitude of energies at the single molecule level is the thermal energy: $k_B T = 4.1 \times 10^{-21}$ J $= 4.1$ pN $\cdot$ nm at room temperatures (where k_B stands for Boltzmann constant and T is the temperature in Kelvin). This energy scale allows one to judge the relative stability of various bonds, whose energies will usually be expressed as multiple of this "quanta". Weak (hydrogen) bonds are typically characterized by energies of a few $k_B T$, whereas the energy of a typical covalent link is 100 $k_B T$. The hydrolysis of ATP, the universal source of energy in the cell, is about 20 $k_B T$. The energy required to separate two complementary bases along a DNA molecule is of order 2 $k_B T$.

3.2 Distances

Two length-scales intervene in typical DNA micromanipulation experiments. First, the length-scale associated to DNA. It can be the distance between successive bases along the helical axis (a fraction of a nanometer), the persistence length of the DNA molecule (about 50 nm), or the end-to-end distance of DNA molecules, which is determined by the number of the base pairs of the molecule, a parameter fixed by the experimenter which usually varies from hundred nm to tens of μm. The second length-scale is determined by the dimensions (typically 1 μm) of the sensor used to exert and/or measure forces on the molecule.

3.3 Times

The dimensions of bio-molecules are so small that inertial effects are completely negligible. Their dynamics is completely dominated by dissipation (viscosity). Biological timescales vary over many orders of magnitudes: electronic dynamics in molecules occurs on psec timescales, whereas protein conformational dynamics occurs on timescales varying from ns to msec. Elementary biochemical reactions vary from 1 μs to 1 ms. Enzymatic cycles may take msec to seconds. While fluorescence energy transfer methods (see previous chapter) allow to study the whole dynamical spectrum of dynamics, finite frequency bandwidth of single molecule manipulations (typically a few tens of Hz) sets a limit on the type of biological phenomena they allow to study, typically enzymatic cycles and their sub-steps.

3.4 Forces

Langevin's force. The small size of the molecular motors makes then extremely sensitive to shocks caused by the thermal agitation of water molecules. The collision of each molecule of water on a protein produces a large instantaneous force, but as the protein experiences as many collisions with molecules coming from one direction as from its opposite, the average force is zero. On short enough timescales, however, the statistical fluctuations on the number of collisions may be so large that the exerted force does not completely cancel. The net fluctuating force exerted by thermal shock thus depends crucially on the observation time and the size of the particle experiencing the collisions. Typically, for measurements on the order of seconds on a bead size of $1\,\mu$m, the force due to thermal shock is about 0.01 pN (this force decreases with the square root of the observation time: the longer one averages the smaller the force due to thermal collisions).

Entropic force. In the absence of tension on DNA, the molecule adopts a random coil configuration in which the local orientation along the molecule is random, this configuration maximizes the entropy of the molecule. When DNA is stretched the number of possible configurations the molecule can adopt, i.e. its entropy is reduced. This reduction in entropy gives rise to a force that resists stretching. It has the same origin as the force resulting from pulling on an elastic string. The free energies involved in this reduction in entropy are on the order of k_BT per degree of freedom. For a DNA molecule modeled as a freely jointed chain with segments on the order of the persistence length $\xi = 50\,\text{nm}$, entropic forces come into play at a force: $F \sim k_BT/\xi \sim 0.1\,\text{pN}$.

Bond breaking force. To get an estimate of the breaking strength of a bond one can use dimensional analysis. The typical force will be given by the ratio of the bond energy to its size. For a weak (hydrogen) bond such as the one between complementary bases in DNA, the energies are typically $E_{gf} \sim 2\text{–}3k_BT$ and the associated dimensions is the radius r of the molecule, about 1 nm. Hence the force required to unpair two bases is approximately: $F \sim E_{gf}/r \sim 10\,\text{pN}$.

Another interesting bond is the noncovalent link between biotin and streptavidin. That receptor-ligand pair is used in many single molecule manipulation experiments to anchor biotin-labelled DNA to a surface coated with streptavidin. At 160 pN it is one of the strongest noncovalent bonds and sets an upper limit on the forces that can be applied to a DNA molecule in these type of experiments. At larger forces the link binding DNA to the manipulation surface is broken.

Finally, the strongest bonds at the molecular level are covalent (interatomic) bonds. The energies involved are on the order of eV (i.e. about $40\,k_BT$) and the distances on the order of 0.1 nm. The tensile strength of a covalent bond is thus on the order 1 nN. Notice however that these order of magnitude estimates do not provide a satisfactory description of the dynamical phenomena that underlies bond breaking, since any bond subjected to a force will eventually break (as a result of thermal fluctuations).

4 Single molecule manipulation techniques

Various micromanipulation techniques have been developed in the past twenty years primarily to manipulate single DNA molecules, but also other fibers, such as actin, microtubules and study their associated molecular motors. Here we will present three of these methods and compare their performance: microfibers, optical tweezers and magnetic traps. The general principle of these methods is the following: a DNA molecule is anchored at one end to a fixed surface and at the other to a force sensor (bead or microfiber). The displacement of the surface or of magnets positioned above the sample creates a force that is sensed by the sensor.

4.1 Optical microfiber

In this technique, DNA is anchored at one end to a microbead held in a micropipette and at the other to a microfiber (Fig. 3.5). The microfiber, often an optical fiber with a core that has been reduced by etching, has a typical diameter of 1 μm and is between 1 mm and 1 cm long. As the micropipette is moved it generates a force on the DNA which is transmitted to the fiber which bends. The extant of bending is detected by the change in the exit position of a laser beam in the fiber (imaged for example on a camera). From the known (or measured) bending rigidity of the fiber (typically on the order of 10^{-6}–10^{-2} N/m) and its deflection one can deduce the force exerted on the DNA. This method provides a spatial resolution on the order of 10 nm and can measure forces down to a few PicoNewtons.

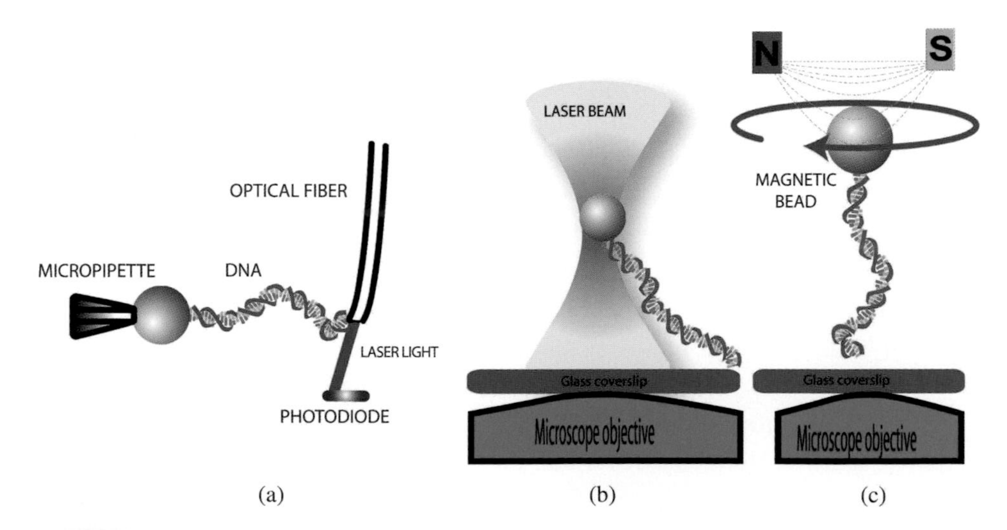

Figure 3.5. Sketch of various manipulation schemes. (a) optical microfibers, (b) Optical tweezers, (c) Magnetic traps. (Figure inspired by Ref. [4].)

4.2 Optical tweezers

Optical tweezers use the force exerted by light on transparent objects to manipulate them. Hence the first step in optical trapping is to pass a laser beam through a high numerical aperture lens (Fig. 3.5). The resulting strong light intensity gradient at the focus balances the radiation pressure exerted on transparent particles (typically 1 μm polystyrene or glass beads in water). It is thus possible to trap such beads near the focal point of a laser beam and manipulate them by translating the beam. By tethering the beads to a surface via a DNA molecule, one can pull on that tether by displacing the surface and measure the force on the DNA by measuring the displacement of the bead in the optical trap. Optical tweezers are naturally extension clamp micromanipulation devices: the experimenter sets the position of the optical trap and the force on the trap beads results from the equilibrium between the optical trapping force and the tension in the tethering molecule. Notice that by an appropriate feedback loop one can transform that set-up into a force clamp device (see below). The stiffness of the sensor is governed by the interaction potential between the bead and the beam. It can be simply tuned as it is proportional to the intensity of the laser beam. The typical values used in experiments on DNA range from 10^{-5} to 10^{-3} N/m. One should nevertheless always calibrate the stiffness of the trap (it can be different for different beads) before any measurement of the bead displacement in order to deduce the force. In most cases, optical tweezers cannot block (or induce) the rotation of the trapped bead. However several ingenious methods have been developed to apply a torque with optical tweezers. The first uses absorbing or birefringent particles which are sensitive to the angular momentum of the trapping beam and can thus be rotated or used to apply a torque. The second uses an anisotropic beam, generated by the higher modes of a Gaussian beam, or by some asymmetrical interference pattern, to create an anisotropic optical trap that can be used to apply a torque on the bead. Finally a third possibility is to use anisotropic particles.

4.3 Magnetic traps

The principle of magnetic traps is simple: a DNA molecule is used to anchor a superparamagnetic bead (of dimension $r \sim 1$ μm) to a surface placed on an inverted microscope (Fig. 3.6). A pair of permanent magnets, stationed above the sample, generates a strong magnetic field gradient on the sample. The tethered bead is thus subjected to both a vertical force (along the field gradient) and a torque that results from the coupling between the beads magnetic moment and the (horizontal) direction of the field. The poles of the magnets are separated by a fraction of a mm, which sets the scale of variation of the magnetic field and its associated force.

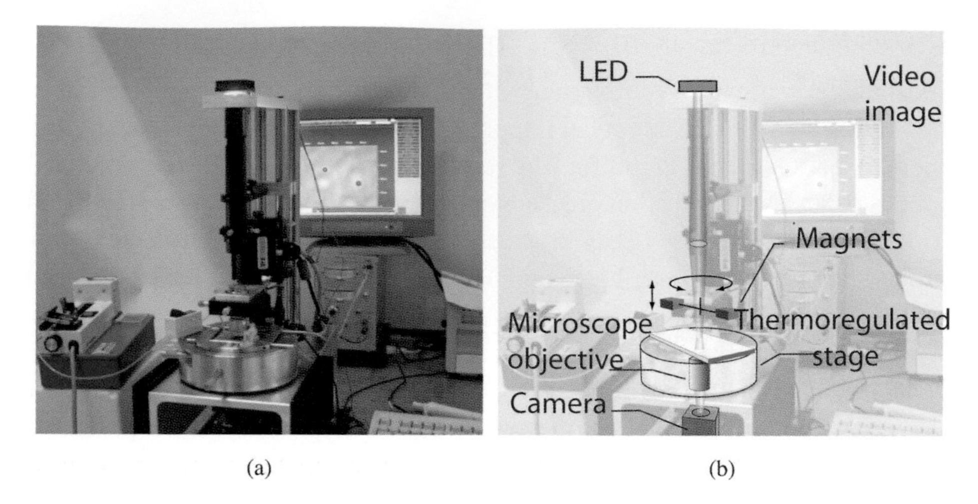

(a) (b)

Figure 3.6. The magnetic trap set-up. (a) Picture of the whole device. (b) Principle scheme: a LED used as light source is focused on the sample and the illuminated area is imaged onto a CCD camera. Magnets placed above the sample exert a force and a torque on magnetic beads anchored to the surface by a single DNA molecule (red). The bead, the DNA molecule and the magnets are not shown at their proper scale.

At the scale of the bead's displacements (typically 1 μm) the applied force is thus essentially constant. It can be controlled by setting the distance of the magnets to the sample.

That system therefore differs from optical tweezers in its principle: it is naturally a force clamp rather than an extension clamp device. The force is set by the experimenter (by setting the distance of the magnets to the sample) and it is the extension of the molecule that relaxes to its equilibrium value. A torsional constraint can be easily imposed by simply rotating the magnets, which induces a rotation of the bead just as the needle of a compass follows the direction of the magnetic field. A further advantage of the magnetic traps is that they pull simultaneously on all the beads in the sample with similar force. Thus they allow multiple parallel measurements on single molecules. Microfibers can only pull on one molecule at a time and optical tweezers are also limited in the number of beads they can trap simultaneously. The beads' positions are recorded in real time (at a typical bandwidth of 60 Hz) with a resolution on the order of 1 nm. The force can be deduced from the transverse Brownian fluctuations of the beads (as described in greater details below). The sensor (i.e. the magnetic bead) imposes no stiffness: indeed, whatever its displacement the force acting on it is the same. In fact, the stiffness is determined by the elasticity of the molecule anchoring the bead. This value depends on the tension in the molecule. For a double stranded DNA molecule of length 1 μm under a tension of 0.5 pN, the stiffness is about 10^{-6} N/m. That stiffness increases significantly with the tension:

at 10 pN, it is approximately 10^{-3} N/m. That stiffness controls the amplitude of the bead's vertical fluctuations and thus the noise in the measurement of the molecule's extension.

5 Comparison of the different techniques

5.1 Qualitative comparison

Micromanipulation techniques can be divided into two groups:

— Extension clamp methods, that control the extension of the molecule and measure the resulting force (microfibers, optical tweezers) and
— Force clamp methods that fix the force and measure the resulting extension (magnetic traps).

As it is often more convenient to study biological system under constant force, magnetic traps are advantageous (even though optical tweezers can be made to work at constant force by an appropriate feedback). Moreover, magnetic traps allow for an easy application of a torsional constraint on the molecule. Notice however that while it is easy to set the degree of supercoiling of a DNA molecule (by rotating the magnets by a set number of turns), it is much more difficult to control the torque on the DNA (since the interaction of even the Earth magnetic field with the bead yields a torque larger by orders of magnitude than the relevant torques on DNA, i.e. a few $k_B T$).

5.2 Comparative performance of the different systems

Force resolution limits. As noted above, the ultimate force resolution of any micromanipulation method is determined by the strength of the Langevin (thermally induced) force. This force depends on the viscosity of the medium (water), the measurement bandwidth (which is related to the rate of the observed phenomena (typically between 1 and 100 Hz); the slower they are, the more one can average, i.e. reduce the bandwidth) and the size of the measured object (i.e. the size of the sensor). Since the dynamics of the observed system is often not under the control of the experimenter, the performance of different manipulation techniques are ultimately set by the single available control parameter: the size of the force sensor.

In the case of a microfibre, the resolution in force is typically a few pN for a bandwidth of the order of a few hundred Hz, but can go down to a fraction of pN by reducing the bandwidth to a few Hz. For optical tweezers, the typical force resolution is 0.1 pN (for a bead of 1 μm and a measurement bandwidth of 100 Hz). In the case of magnetic tweezers the force is usually calibrated before any measurement and it can be measured with abitrary precision. The limit in this case is practical: to obtain an accuracy of the order of 10% at 0.01 pN requires an acquisition of a few tens of minutes. This technique is however unable to precisely measure high forces (see discussion in next paragraph). The maximum force measured for a bead of diameter 1 μm attached to a DNA molecule of length 1 μm is typically 10 pN for a bandwidth of 60 Hz (one can achieve larger forces using bigger magnetic beads).

Distance resolution limits. The theoretical resolution limit of optical microscopy methods (such as used in micromanipulation techniques) is of the order of δ 200 nm. This value sets a bound on the ability to separate two nearby objects, but it does not limit the accuracy of position measurement for a single object. Indeed the position of an object (bead, fluorophore, etc.) can be measured with arbitrary accuracy ($\delta/\sqrt{N}$), where N is the number of collected photons arriving from the object. In practice, due to the saturation of CCD pixels, particle tracking methods generally allow to track the position of a 1 μm particle with a precision of about one nanometer (at typical video rates). However, the main limiting factor in spatial resolution is often imposed by the Brownian fluctuations of the sensor which is connected via the stiffness of the trap (or of the tethering molecule k) to fluctuations in the thermal Langevin force mentioned before.

Summary of the characteristics of the three micro-manipulation methods.

The table below summarizes the major characteristics of the three micromanipulation set-ups presented above, namely the range of available forces, their spatial resolution and their stiffness.

Box 3.1. Summary table of the performances of the three methods.

Set-up	Force range (pN)	Spatial resolution (nm)	Sensor stiffness ($pN-nm^{-1}$)	Bandwidth (Hz)
Optical tweezers	0,1–100	0,1–5	5×10^{-3}–1	50–5 000
Magnetic traps	0,001–100 000	2–5	10^{-6}	10–1 000
Microfibres	10–1 000	10	2×10^{-3}–1	50–200
AFM	5–1 000	0,1–1	1–10^{5}	1 000

6 DNA mechanical properties

It is often taught that chemical bonds make well defined angles. However this does not take into account the thermal agitation which adds noise to the average bond angle, not to mention possible rotations around these bonds. For a macromolecule such as DNA, with a large number of successive links, this thermal noise results in a decorrelation of the molecule's orientation over a distance known as the persistence length.

DNA is a very interesting polymer. First, it is much more rigid than most synthetic polymers: its persistence length is approximately $\xi = 50\,\text{nm}$, while it is only 1–2 nm for polymers such as poly-ethylene or poly-vinyl chloride (PVC). Second, as a consequence of its double helical structure, DNA opposes twisting: it has a non-zero torsional modulus. This is in contrast with synthetic polymers or single strand DNA which are free to rotate about their chemical bonds.

As a result of its large persistence length DNA is an ideal polymer, that is to say a polymer where self-avoiding interaction can be neglected for molecules up to several tens of microns in length. This has allowed the use of DNA as an experimental model for the study of the mechanical behavior of an ideal polymer, the only model that can be solved exactly.

The simplest model of such a polymer is the freely-jointed-chain (FJC) model, which maps the conformation of a polymer of size L on a random walk of $N = L/b$ steps (where $b = 2\xi$ is known as the Kuhn length). Such a polymer forms a random coil of mean square radius: $R_g = \sqrt{\langle R^2 \rangle} = \sqrt{bL}$. If the polymer is pulled with a force F, the random walk is biased in the direction of the force and the extension of the polymer (the distance l between its extremities) is obtained as a balance between the work Fl done against the entropy and the free energy of the polymer, which at low forces increases as: $3k_BTl^2/bL$. For the mean distance between the polymer's ends to be l a force $F = 3k_BTl/bL$ has to be exerted. This linear relationship between force and extension means that at low extension an ideal polymer such as DNA should behave as a Hookean spring. As the spring stiffness (the factor in front of the l-dependence) depends on temperature one speaks of an entropic spring. Notice that the stiffness increases with raising of temperature, resulting in a contraction of the polymer (at fixed force). That is in contrast with non-entropic springs which stiffness usually decrease with temperature.

While this model is a valid approximation at low forces $F < k_B/b \sim 0.08\,\text{pN}$ (i.e. at small extensions: $l < L$), DNA stretching experiments have shown that the FJC model does not describe its behavior at larger forces. In this regime, DNA behaves not as a chain of freely rotating rigid segments, but as a more realistic flexible tube. This model, known as the Worm-like-Chain (WLC) model is also exactly soluble and its predictions fit the experimental data extremely well (see Fig. 3.7) up to forces of about 10 pN. At larger forces, DNA is fully extended and stretches like a regular spring: F = S $\delta L/L$ with $S \sim 1$ nN. This regime extends up to a force of about 60 pN, at which point DNA undergoes a reversible transition to a new structure

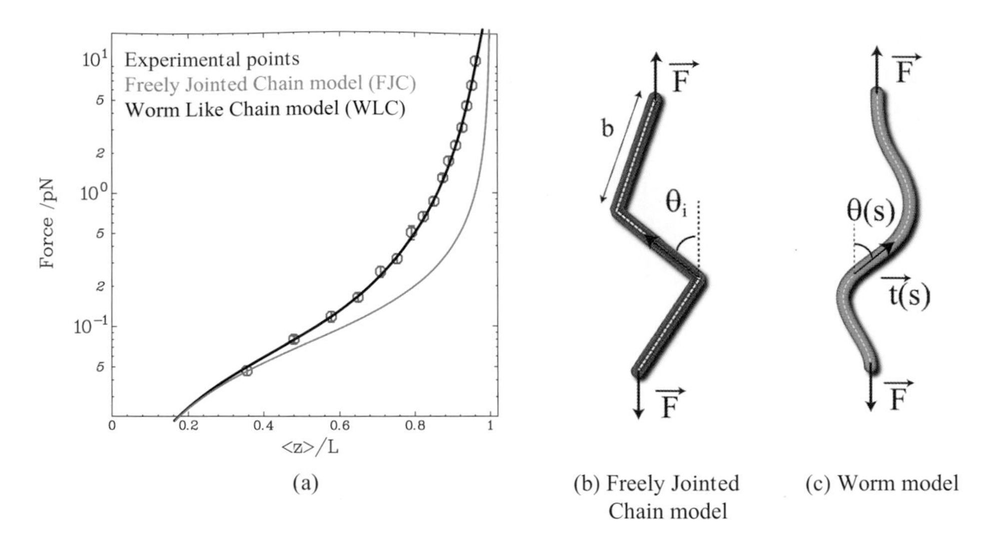

Figure 3.7. Mean relative extension $\langle z \rangle / L$ of a DNA molecule under tension F. Notice the excellent agreement with the theoretical prediction of the WLC model and the agreement at low forces with the FJC model (which makes similar predictions as the WLC model in this low force regime). (Figure inspired from Ref. [6].)

called S-DNA, which is ~70% longer than B-DNA, the classic form discovered by Watson and Crick. S-DNA is an unstable structure: in the presence of a nick in one of its two strands, they may separate.

As mentioned above, one of the characteristics of DNA, which singles it out from synthetic polymers, is its resistance to torsional stress. The magnetic trap set-up described above offers a simple way to twist the molecule: just rotate the magnets that pull on a magnetic bead tethered to the surface by a single DNA molecule (notice, however, that the DNA must not be nicked and should be anchored at more than one point to the surface and the bead to prevent relaxation of torsional stresses by rotation about one of the strands). The investigations of single DNA molecules under tensional and torsional stresses are very instructive. There results can be intuited from our daily experience with twisted cords or tubes. Thus if one twists a DNA molecule under tension clockwise or counter-clockwise, it responds symmetrically by shortening. Beyond a certain torsional stress it forms tertiary supercoiled structures, known as plectonemes, which cause an increased shortening of the end-to-end distance of the molecule (Fig. 3.8). The essence of these observations is captured by a simple calculation. A DNA molecule of extension l under tension F and twisted by n turns experiences (like an elastic tube) a torque Γ given by:

$$\Gamma = 2\pi nC/L$$

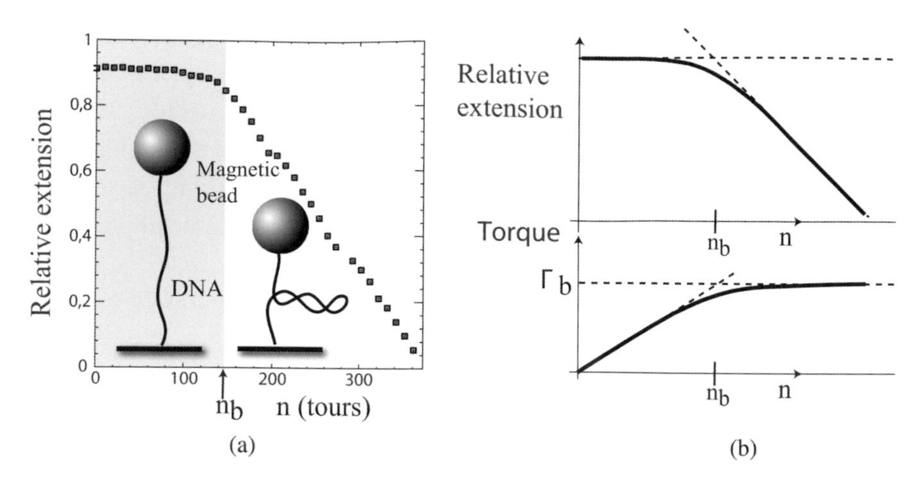

Figure 3.8. Mean relative extension $\langle z \rangle / L$ of a DNA molecule under a tension F as a function of the number of turns n. (a) Passed a certain threshold n_b, the molecule contracts by forming tertiary supercoil structures known as plectonemes. (b) Prediction of a simple mechanical model of buckling in DNA. The torque in the molecule Γ grows until the buckling threshold is reached and remains constant thereafter.

where $C = k_B T \mathscr{C}$ is the DNA torsional modulus ($\mathscr{C} = 100\,\text{nm}$). At small n, the molecule responds by twisting slightly with only a slight decrease in its extension. As the torsional energy increases as $\pi n \Gamma$ i.e. quadratically with n, past a certain number of turns n_b (and associated torque Γ_b) it becomes energetically less costly for DNA to buckle (as a tube would) and form a loop rather than for its torsional energy to increase. The formation of a loop involves two energy terms. First, the bending energy of the loop: $E_b = 2\pi R(B/2R^2)$ (where $B = k_B T \xi$ is the DNA bending modulus). Second the work performed against the tension pulling on the molecule to form the loop: $W = 2\pi RF$. Balancing that energy against the torsional work yields:

$$2\pi \Gamma_b = E_b + W = \frac{\pi B}{R} + 2\pi RF.$$

The radius of the loop that minimizes the energy term on the right is given by: $R = \sqrt{B/2F}$, so that the critical buckling torque becomes: $\Gamma_b = \sqrt{2BF}$ and the number of turns at buckling:

$$n_b = \frac{L\sqrt{2BF}}{2\pi C}.$$

Thus, the greater the force, the greater the torque Γ_b, the larger the number of turns until buckling n_b and the smaller the radius of the loop at buckling. Past that buckling transition, as the molecule is twisted it continues to wind around itself, but the torque remains constant. Therefore increasing the number of turns on the molecule leads to an increase in the number of supercoils (plectonemes) and a proportional reduction in the end-to-end distance of the molecule. This model provides a semi-quantitative description of the behaviour of DNA under torsion. While it predicts a sudden buckling transition as a critical torque is reached, in practice thermal fluctuations, which are not accounted for in this simple estimate, soften the transition especially at low forces.

The previous model treats DNA as an elastic tube under torsion, however DNA possesses a chiral helical structure which is sensitive to the torsion in the molecule. In particular upon unwinding (negative twist) if the torque Γ reaches a threshold $\Gamma_d \sim 9\,\text{pN nm} < \Gamma_b$, the molecule denatures locally: it responds to the torsion by an unwinding of the two strands by about 10.5 pairs per additional turn (Fig. 3.9). To observe this distortion with a molecule under tension, the force must be large enough so that the critical buckling torque is higher than the critical torque for

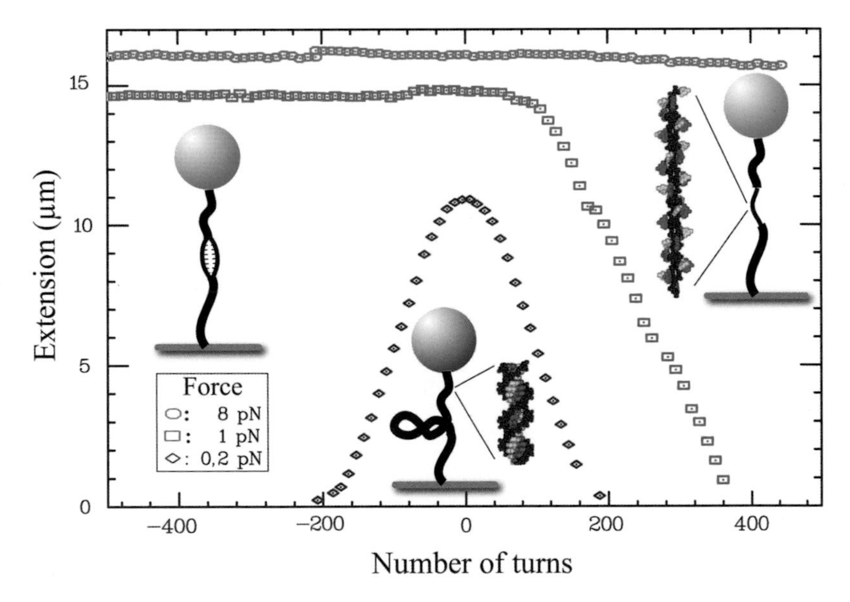

Figure 3.9. Mean relative extension $\langle z \rangle / L$ of a DNA molecule as a function of the number of turns n at various tensions F. At small forces $F < 0.5$ pN, as the molecule buckles under positive or negative coiling, the extension is symmetric under $n \to -n$. At intermediate forces $0.5 < F < 3$ pN, the extension is asymmetric with a significant decrease at positive supercoilings where the molecule buckles and a rather constant extension at negative supercoils (as the molecule denatures rather than buckles). Finally at larger forces $F > 3$ pN, the extension under postive supercoiling varies little, as the DNA molecule adopts a novel structure, P-DNA, characterized by a shorter pitch (the phosphate backbones winding around each other at the core of the molecule and the bases exposed in solution), see inset: DNA structure obtained by R. Lavery from a numerical simulation of DNA under positive torque.

denaturation $\Gamma_b > \Gamma_b$. In practice, this condition is satisfied for a tension greater than about 0.5 pN. Thus, for forces greater than this threshold, the response of DNA to twist is asymmetric: it displays little change in extension upon negative winding (as it denatures), but decreases significantly (by about 50 nm per turn) for positive winding as it forms plectonemes.

This is true up to a force of about 3 pN, beyond which DNA does not supercoil anymore when positively twisted. At these forces, the positive torque on the molecule is sufficiently large to induce another structural transition, to a compact and highly twisted DNA structure, called P-DNA. This structure is characterized by a right-handed double helix about 75% longer than B-DNA and a helical pitch 4 times smaller (2.6 bases per turn). The phosphate backbones of this new structure of DNA are wound around each other at the core of the molecule while the bases are exposed to the outside. Such a structure could only be discovered by single molecule manipulations of DNA where both the tension and the torsion in the molecule can be controlled.

6.1 Mechanical properties of single stranded DNA

While double-stranded DNA was used to test different mechanical models of an ideal polymer (sensitive or not to torsional stress), the elastic behavior of single-stranded DNA (ssDNA) is much more complex. ssDNA (just as RNA) is a very flexible polymer with a persistence length of about 2 nm. The charges on its phosphate backbone plays an important role in the molecule's self avoidance by generating an electrostatic repulsion field (screened at a distance of a few nm) between its different parts. Moreover the possibility of base-pairing between complementary bases along the molecule adds a further degree of complexity by allowing partial hybridization between complementary regions along the molecule. However, by chemically modifying the bases along ssDNA, one can obtain a very long polymer (a few microns) in which only the self-avoidance interactions (in addition to entropy) control the elastic behaviour. The measured response of such a polymer to tension agrees remarkably well with simple dimensional arguments (described below) for how such polymer should behave.

For a self-avoiding polymer chain, the characteristic size of the random coil it forms is given by the Fleury radius: $R_F = \sqrt{\langle R^2 \rangle} = L^{3/5}(v/b)^{1/5}$ (where v is the excluded volume). To describe the behaviour of such a polymer under tension, one assumes that the only characteristic dimension controling that behaviour is R_F. In that case one can describe the polymer extension l under a tension F as:

$$l = R_F \Psi\,(F \times R_F / k_B T).$$

where $\Psi(x)$ is some unknown function. At low forces, i.e. $x = FR_F/k_BT \ll 1$ one expects the polymer to behave as an entropic spring, i.e. $\Psi(x) \sim x$. At intermediate forces, but still in a regime where entropy is relevant, i.e. for $F < F_c = k_BT/b \sim$ 2 pN, one expects the extension l of the polymer to be some fraction of its total length L. To satisfy that relation in that regime $\Psi(x) \sim x^{2/3}$, which then yields:

$$l/L = (v/b^3)^{1/3}(Fb/k_BT)^{2/3}.$$

This behaviour should be valid for extensions $l < l_c = L(v/b^3)^{1/3}$. As the excluded volume v increases at low salt (as the Debye length increases), one can compare the behaviour of ssDNA at different salt concentrations and test the simple dimensional analysis described above. Renormalizing the lengthscales by l_c and the force-scales by F_c, these dimensional arguments predict a universal elastic behaviour (independent of environmental conditions) which at low forces vary as $l/l_c = (F/F_c)^{2/3}$. This indeed is what is observed experimentally (Fig. 3.10). It is interesting to notice that at forces above F_c, the behaviour is also universal but the relative extension then increases as $\log(F/F_c)$. This behaviour which has also been observed for denatured proteins has so far no explanation.

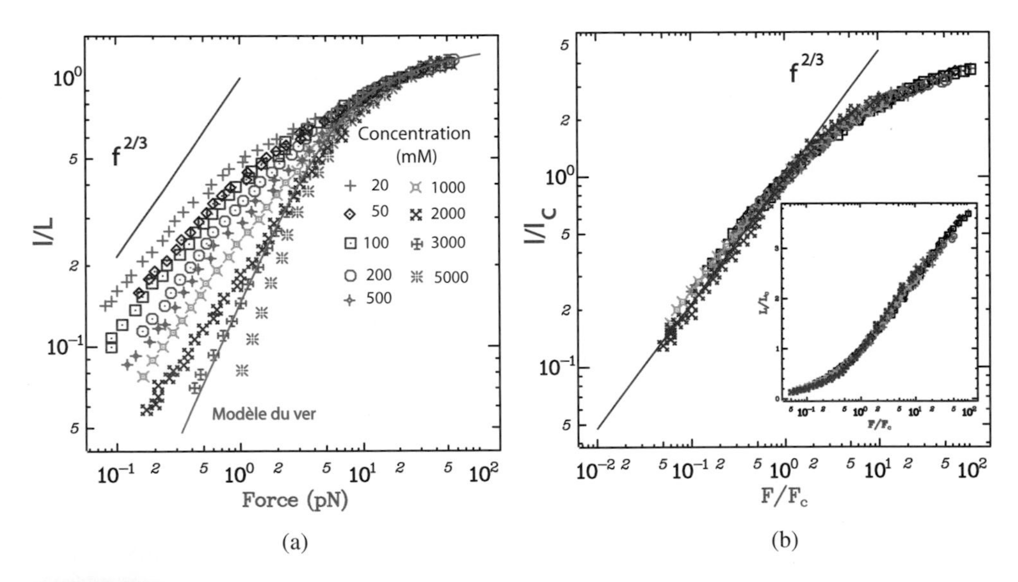

Figure 3.10. (a) The mean relative extension l/L of an ssDNA molecule under tension F at different salt concentrations. (b) The same data renormalized by the critical force and length scales: F_c and l_c. Notice that all curves collapse into a single curve with a relative extension that at low forces l/l_c varies with the tension as $(F/F_c)^{2/3}$, as predicted. As can be seen in the inset on the right, at forces larger than F_c the extension increases as $\log(F/F_c)$, a behaviour that is still unexplained. Data are from Ref. [7] (data supplied by O.A. Saleh, UCSB).

7 Conclusion

Single molecule micromanipulation experiments allows for testing the elastic behaviour of polymers such as DNA (but also other biological fibers such as actin or microtubules) which are important from a fundamental physical and biological point of view. Knowledge of these properties can allow one to monitor the biological processes involving these polymers, since these often affect their mechanical properties. The micro-manipulation tools described in this chapter can therefore be used to shed new light on biological processes involving DNA. In Chapter 4 we will see how these techniques can be used to study molecular motors, the energy-driven motion of proteins on DNA and other fibers. These techniques can also be used to study the interaction of structural proteins with DNA since the force can alter their binding affinity or the distance between various elements. These approaches have also found a variety of other applications. Thus micromanipulation experiments have been used to study how proteins fold and unfold when put under tension. Over the past twenty years, micro-manipulation techniques have thus become an important addition to the toolbox used in the study of biological systems.

Bibliography

[1] The web-page of the group of S.Block has much data on optical tweezers and various molecular motors: http://www.stanford.edu/group/blocklab/. In particular see the two papers below.

[2] Abbondanzieri EA, Greenleaf WJ, Shaevitz JW, Landick R, Block SM. Direct observation of base-pair stepping by RNA polymerase. *Nature* (2005) **438**, 460–465.

[3] Neuman KC, Block SM, Optical trapping. *Rev. Sci. Instrum.* (2004) **75**(9), 2787–809.

[4] Neuman, Liunnet T, Allemand JF. Single-molecule micromanipulation techniques. *Ann. Rev. Mater. Res.* (2007) **37**, 33–67.

[5] Gosse C, Croquette V. Magnetic tweezers: micromanipulation and force measurement at the molecular level. *Biophys. J.* (2002) **82**, 3314–3329.

[6] Charvin G, Allemand JF, Strick TR, Bensimon D, Croquette V. Twisting DNA: single molecule studies. *Contemp. Phys.* (2004) **45**, 383–403.

[7] Saleh OA, McIntosh DB, Pincus P, Ribeck N. Nonlinear low-force elasticity of single-stranded DNA molecules. *Phys. Rev. Lett.* (2009) **102**, 068301.

4

Molecular motors

Jean François Allemand, *Professor at the Ecole Normale Supérieure, Laboratoire de Physique Statistique de l'Ecole Normale Supérieure, Paris.*
Pierre Desbiolles, *Professor at Pierre and Marie Curie University, Kastler Brossel Laboratory, Paris.*

How do we move? More precisely, what are the molecular mechanisms that can explain that our muscles, made of very small components can move at a macrosopic scale? To answer these questions we must introduce molecular motors. Those motors are proteins, or small protein assemblies that, in our cells, transform chemical energy into mechanical work. Then, like we could do for a macroscopic motor, used in a car or in a fan, we are going to study the basic behavior of these molecular machines, present what are their energy sources, calculate their power, their yield. If molecular motors are crucial for our macroscopic movements, we are going to see that they are also essential to cellular transport and that considering the activity of some enzymes as molecular motors bring some interesting new insights on their activity.

1 A rotating motor: ATP synthase

We are first focusing on a rotary motor called ATP synthase (Fig. 4.1). This motor drives the synthesis of a small molecule, adenosine tri phosphate (ATP), which is the

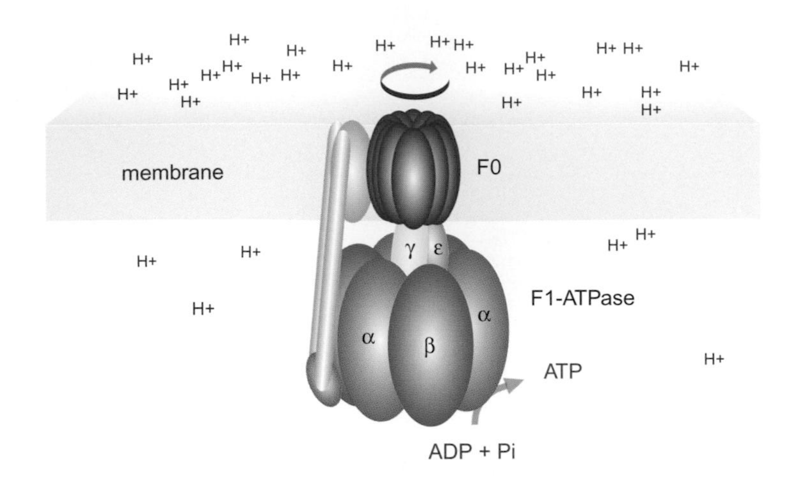

Figure 4.1. ATP synthase is built from two components named F0 and F1 (or F1-ATPase). A proton gradient between two sides of a membrane in which the motor part F0 is anchored produces the rotation of a central shaft (γ subunit) linking F0 to F1. This allows ATP synthesis from ADP and inorganic phosphate Pi.

main chemical energy source of our cells. The hydrolysis of a single ATP molecule releases a molecule of adenosine diphosphate (ADP) and a phosphate group (Pi): $\mathrm{ATP} \rightarrow \mathrm{ADP} + \mathrm{Pi}$ but also a tremendous energy of about $20\, k_{\mathrm{B}}T$. This reaction is reversible if one can provide the energy to reform the chemical bond between ADP and Pi. This is exactly what ATP synthase is doing, in a cellular compartment called mitochondria (see Chapter 1). To understand how this motor operates let us first consider its structure, at least the main elements that are present from bacteria to Humans. ATP synthase is a protein complex made of two principal parts named respectively F0 and F1 (Fig. 4.1). Its operating principle can be summarized in the following way: a difference in the proton concentration across the mitochondria membrane in which the F0 part in engaged produces the rotation of F0 and of the shaft that links F0 and F1. This rotation produces structural changes in the F1 part. The mechanical energy to deform F1 produces the energy to generate the bond between ADP and Pi if those molecules are bound to the F1 part in the proper conformation and then syntheses ATP.

The F0 part is quite difficult to manipulate since as it is engaged in a membrane, it has highly hydrophobic parts. The F1 part, also named F1-ATPase is easier to purify since, as it lies inside the mitochondria, it is hydrophilic. Many theoretical and experimental work were dedicated to the understanding of the coupling between mechanical rotation and chemical synthesis. In particular Boyer proposed the basic mechanism for this coupling and Walker obtained F1-ATPase structure. For this work both obtained the Nobel Price in Chemistry in 1997.

1.1 ATP hydrolysis produces the rotation of F1-ATPase

As we have seen the main role of ATP synthase is to produce ATP from the electrochemical energy of a proton gradient concentration. Nevertheless it can also work in the opposite way: F1-ATPase can hydrolyze ATP to change its structure and to induce the rotation of the central shaft linked to the F0 part.

The pioneering experiment demonstrating directly that F1-ATPase could use ATP energy to rotate was performed by the group of Prof. Kinosita in Japan. Its principle is quite simple: F1-ATPase was attached to a glass surface and shaft rotation was simply monitored by fluorescence microscopy (see Chapter 2). The clever point was to attach a large molecule to the shaft: a short actin filament (Fig. 4.2) that was long enough to provide, when fluorescently labeled, a large signal easily observable under a microscope but small enough to be considered as a rigid beam and then a good tag for the rotation (see Chapter 5). With this experiment Kinosita's team could measure the speed of the motor as a function of ATP concentration, yet the simple observation of the rotation confirmed the proposed models for the working principle of ATP synthase! In addition the researchers have shown that the rotation speed is reduced when the size of the actin filament increases. Indeed hydrodynamics predicts that the drag exerted by water depends on the size of the filament: the longer the filament the stronger the friction and the more energy is dissipated into the surrounding solution. Hydrodynamics also predicts that the torque exerted on the filament in these conditions, characterized by a small Reynolds number, is the product of the rotation speed by a factor depending on the length of the filament and on its radius. Rotation speed and geometry of the filament can be easily measured with a microscope, and then, simply by analysing the rotation of the motor it was

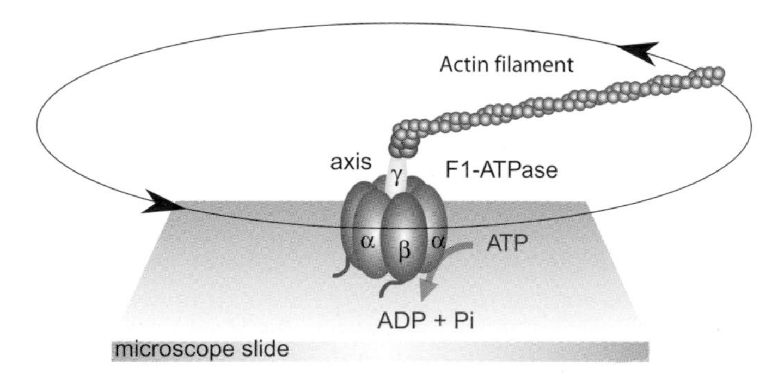

Figure 4.2. F1-ATPase is a reversible motor: energy produced during ATP hydrolysis can rotate γ axis. Binding an actin filament on this axis allows the observation and the measure of the rotation speed of the molecular motor. Figure inspired by Kinosita *et al.* (see Ref. [4]).

possible to measure the torque exerted by the motor. The value obtained by Prof. Kinosita's group was 80 pN · nm.

With these values, the yield of the motor can be calculated provided we have more information on F1-ATPase structure. F1-ATPase can be seen as a barrel, that would be the stator of an electric engine, in the middle of which would pass an axis, named the γ subunit, that would be the rotor of the engine (Fig. 4.1). The barrel itself if made from 6 subunits: 3 α and 3 β subunits, that constitutes 3 $\alpha\beta$ pairs. When rotating, the γ subunit induces conformational changes in the $\alpha\beta$ subunits. When they interact with ATP molecules, or ADP and Pi, bound to some pockets in these subunits, they catalyze chemical reactions on these reactants. These structural elements provide a symmetry for the motor where an $\alpha\beta$ subunit occupies a 120° angle. A synthesis or hydrolysis of an ATP molecule must then occur each time the rotor (γ subunit) rotates by 120° compared to the stator. The energy produced during this 120° rotation is nothing else than the product of this angle by the torque produced by the motor (80 pN · nm). A simple calculation shows that the mechanical energy produced is about 20 $k_B T$, which is almost the chemical energy produced by a single ATP hydrolysis. F1-ATPase is then a tremendously efficient motor: it is able to convert almost all the chemical energy into mechanical energy, in other words its yield is about 100%!

This yield can look surprisingly high if we compare it to the yield of the motors used in our cars, which is at most 40%. But not if one thinks about the electric engines that also have excellent yields, about 90%. Finally, what is maybe the most striking element is that such a small motor, whose size is only 30 nm (the size of F1-ATPase), can rotate a 1 μm actin filament in a fluid in conditions where the Reynolds number associated with the rotation is extremely small. Even if this kind of comparison has its limits, at a macroscopic scale this is equivalent to a 30 cm motor rotating a 10 m tree at a speed of about few tens of turns per second in a medium much more viscous than honey. The almost perfect yield is maybe simply the consequence of a good selection process during evolution since if a poor yield could be accepted for a non reversible motor, it cannot be the case if the motor has to be reversible as it is the case for the F1-ATPase. As we noticed already F1-ATPase can use ATP to rotate its axis, and then in a biological context generate a proton gradient. But its main biological role is to use the rotation of the F0 part resulting from a proton gradient, to rotate the axis of F1-ATPase and then synthesize ATP.

1.2 Producing ATP by rotating F1-ATPase

In fact demonstrating this reversibility in an experiment was not simple. There were two challenges: one not only had to manage the rotation of the axis of the

30 nm motor but also to detect the production of a few ATP molecules. To rotate the γ axis of the motor the solution was to attach it to a magnetic bead and to rotate this microscopic compass with a rotating magnetic field. The remaining challenge was then to show that, in the presence of ADP and Pi, each time the bead was rotated by 120° a single ATP molecule was produced. This is a serious experimental concern: How can we detect the formation of a single molecule? The experiment that we described previously succeeded in relating ATP concentration to the rotation speed of actin filament. So in principle, once some ATP has been produced, and the magnetic field suppressed, the bead attached to the axis should start a rotation generated by the motor in presence of ATP, and by measuring the speed of rotation one should have an indirect measure of ATP concentration. The problem is that this technique works only if the ATP concentration is on the order of a millimolar, that is, 6×10^{20} molecules per liter. But a 360° rotation of the bead does produce only 3 ATP molecules. In a 1 liter volume one would need 10^{20} turns to observe the rotation of the motor in the absence of the magnetic field! The solution was to strongly reduce the volume of solution surrounding the molecular motor, by placing F1-ATPase in micro-reactors (Fig. 4.3). A reactor with a typical dimension of 1 μm has a volume of $10^{-18}\,\mathrm{m}^3$, or 10^{-15} L: only a few thousand turns are needed to reach the required ATP concentration to observe the spontaneous rotation of the axis without a magnetic field. The group of H. Noji did use such a microreactor to quantitatively demonstrate that the motor was reversible and they also demonstrated that a part of the protein complex, named ϵ (Figs. 4.1 and 4.3), is essential for the mechanical synthesis of ATP but not for the rotation produced by its hydrolysis.

1.3 Mechanics and chemistry

The coupling between chemistry and mechanics working in F1-ATPase deserves two other comments, that can be generalized to other biological systems.

First, in the case of the actin filament experiment: What is happening when there are very few ATP molecules in solution? Thanks to its random motion, an ATP molecule in solution will diffuse to the catalytic pocket of the motor where its hydrolysis will produce the rotation of the axis by one third of a turn. Then it will take some time before the next ATP molecule will enter in the next pocket, allowing the rotation to continue. In these conditions where the energy source is rare, one observes a discontinuous rotation of the axis with elementary steps of one third of a turn. For F1-ATPase, like for other molecular motors, the mechanical elementary step is imposed by the structure of the motor. In the case of F1-ATPase the use of a fast camera to monitor the rotation of the axis allowed the experimentalists to distinguish 40° and 80° sub-steps in the 120° steps. Each substeps could be linked

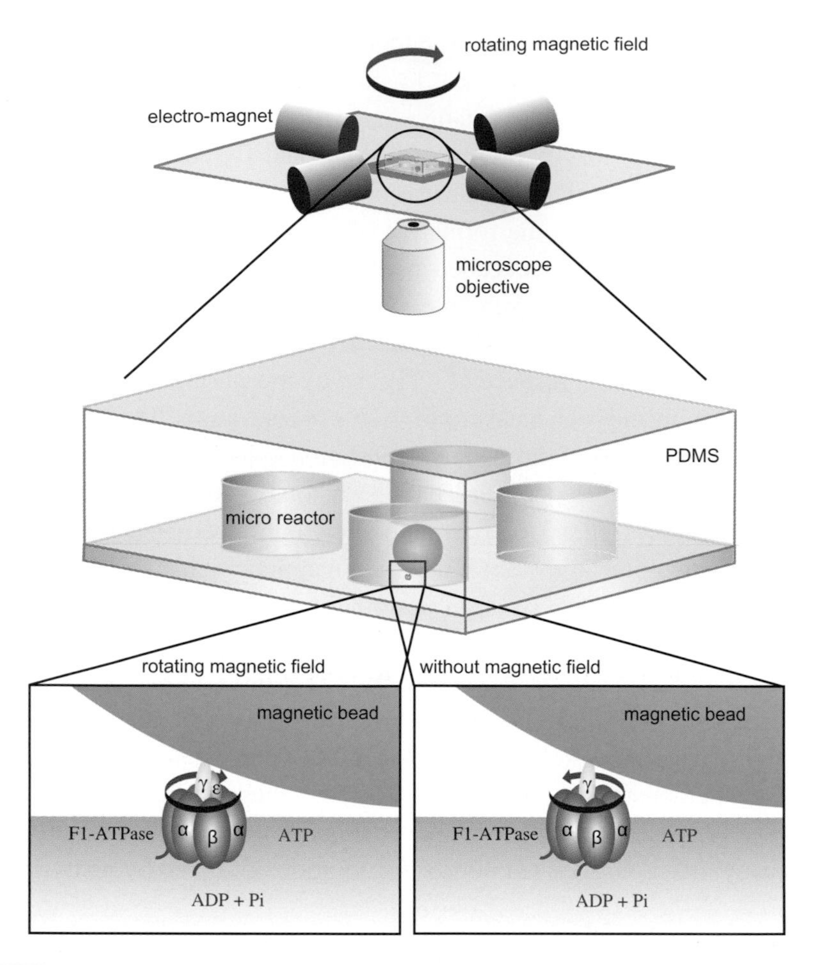

Figure 4.3. The reversibility of the movement of F1-ATPase was illustrated by attaching to the γ axis a magnetic bead manipulated by a rotating magnetic field. The molecular motor and the bead are placed in a PDMS (PolyDiMethylSiloxane is a transparent polymer allowing the observation of the bead with a microscope) microreactor. When the rotating magnetic field is applied to the bead, ATP synthase produces ATP from ADP and Pi present in the reactor. As the volume of the microreactor is tiny, ATP concentration increases very quickly. After a few minutes, the magnetic field is turned off and one observes the spontaneous rotation of the bead in the opposite direction: F1-ATPase now hydrolyses the ATP in solution and uses the chemical energy produced to rotate the γ axis.

to a stage in the ATP hydrolysis cycle, setting in evidence the very strong coupling between the mechanical rotation of the motor and the evolution of the chemical reaction.

This strong coupling between chemistry and mechanics is the second important point. It can be illustrated by the following observation: sometimes the rotation of the F1-ATPase axis is blocked after the first 40° sub-step for an unknown reason. This

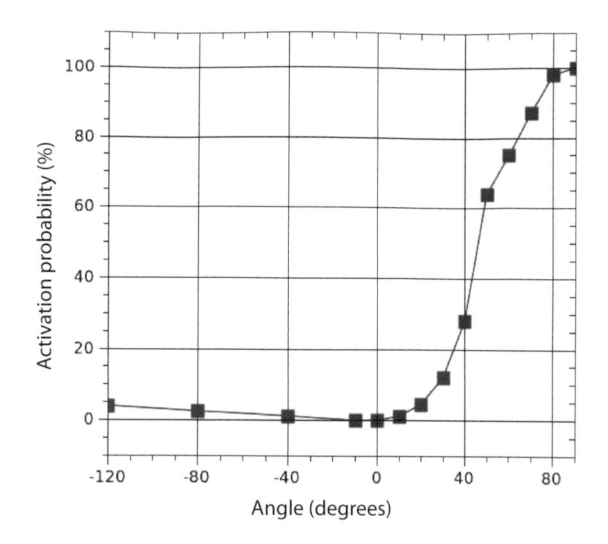

Figure 4.4. Probability of mechanical activation of F1-ATPase as a function of the imposed angle. Inspired from Kinosita *et al.* (see reference [4]).

produces an unwanted stop of the motor. The hypothesis is that the ADP molecule that is in the catalytic pocket cannot be ejected from its binding site. If the link between mechanics and chemistry is real, one can imagine that when forcing the motor rotate it could be possible to eject the blocked ATP and resume the rotation of the motor. The experiment was performed by the group of Prof. Kinosita who used the magnetic bead to force the rotation of the motor as described previously. As the torque produced by the bead is huge at this scale, it can set the angular position of the axis. It is then possible to measure the probability to resume the rotation after being at a given angle after its stalled position. Kinosita's group observed that the motor had a 100% probability to resume its rotation if the rotation angle is about 80° that is missing to go to the next step in the chemical cycle (Fig. 4.4). Mechanics can then help chemistry to end a reaction.

2 Myosins: a linear motor example

We now describe molecular motors which are not involved in rotation but in translation: the myosins. This family of motors contains many members which are crucial for many functions of an organism. When they dysfunction this leads to serious problems such as cardiac diseases in the case of cardiac myosin II. A default in myosin V, used in vesicle transport, can lead to the Griscelli syndrome associated with immune deficiencies, and some mutations of these motors can lead to neurological problems. Myosin IB mutations can lead to deafness (see Chapter 7).

2.1 Translocation on actin filaments

All myosins move on the same track: actin (see Chapters 1 and 5), and all go in the same direction along actin, a polar biopolymer i.e. it has an orientation. Yet myosins may have very different structures and roles. Myosin V, involved in vesicle transport in the cell, has two "feet" so it can "walk" along actin, whereas myosin II, the essential component of our muscles, has a single one. Even though it is not the only reason, the fact that myosin V has two feet makes it easier for it to travel longer distances than myosin II. Myosine V is said to be processive, that is it can make a large number of catalytic cycles (steps) before detaching from its substrate. We must not forget that myosins, just like any molecule, are subject to Brownian motion (see Chapter 3): after each step along actin there is a non zero probability that it detaches from the substrate and then goes away from it. If this probability is large, myosin detaches after a small number of steps, and even after a single step. It is the case for myosin II that has a single attachment point to actin. Its movement along the actin track requires it to detach between two attachment points, and thus the probability to stay bound to actin during a step is obviously zero, just like its processivity. To have a better view of this phenomenon let's take comparison with the human walk: when we walk we always have a foot on the ground, while when we progress by jumping on a single foot, we are not in contact with the ground during the jump. Were we subject to a large Brownian motion during this jumping phase we could not make two steps in the same direction like young kids playing do.

How to determine motor processivity? In the case of myosin, but this approach could be used for other molecular motors, experimentalists designed what is called motility assays (Fig. 4.5). It consist in experiments where motors are attached to a surface with a varying density. The binding is such that the "feet" of the motors are still functional. One then adds a solution with the energy source of the motor, in general ATP, and actin filaments, labeled with fluorescent molecules which allow

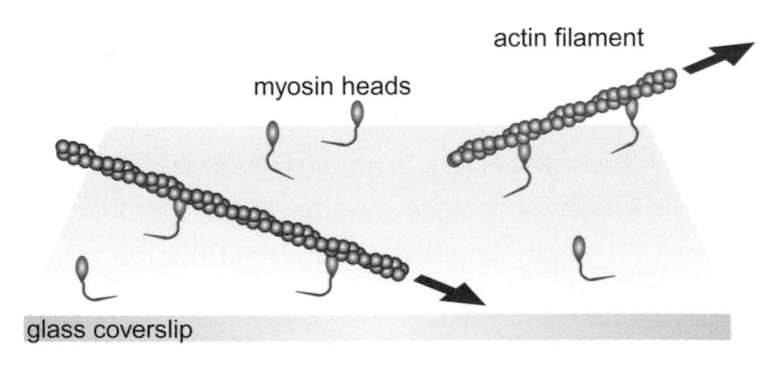

Figure 4.5. Motility assay for myosin: motors are attached to the glass surface and actin filaments move when ATP is added to the solution.

the filament to be imaged with a microscope. When the filaments interact with the motors the experimentalist can see them moving on the surface! By reducing the density of the motor attached to the surface one can reach the regime where the density of motors on the surface is so small that a single motor can interact with a filament. If the motor is processive the filament will move for long distances, which will not be the case if the motor is not processive.

In the case of processive myosins, like myosin V, such single molecule experiments give us more information on their behavior. As we already mentioned, myosin V has two feet, one being always in contact with the actin filament. Two potential ways of moving can be anticipated (Fig. 4.6). Does the motor place one foot after the other like a walker, or, like in fencing, there is always the same foot ahead? Can we know which mechanism is used for myosin V? When a walker travels with a regular step p with a rate of n steps per second, $n/2$ times per second its right foot moves by $2p$. In the other case, the foot moves n times per second (Fig. 4.6) with a step p. Thanks to tremendous advances in fluorescence microscopy, and in particular the spatial superresolution (see Chapter 2) that allows to observe displacements with a nanometric precision, P. Selvin's group in Urbana-Champaign observed that the foot of a single myosin V moves by a distance that is twice the distance of the motor step size. Thus myosin V walks like a standard walker by placing one foot ahead of the other, in a way called hand over hand.

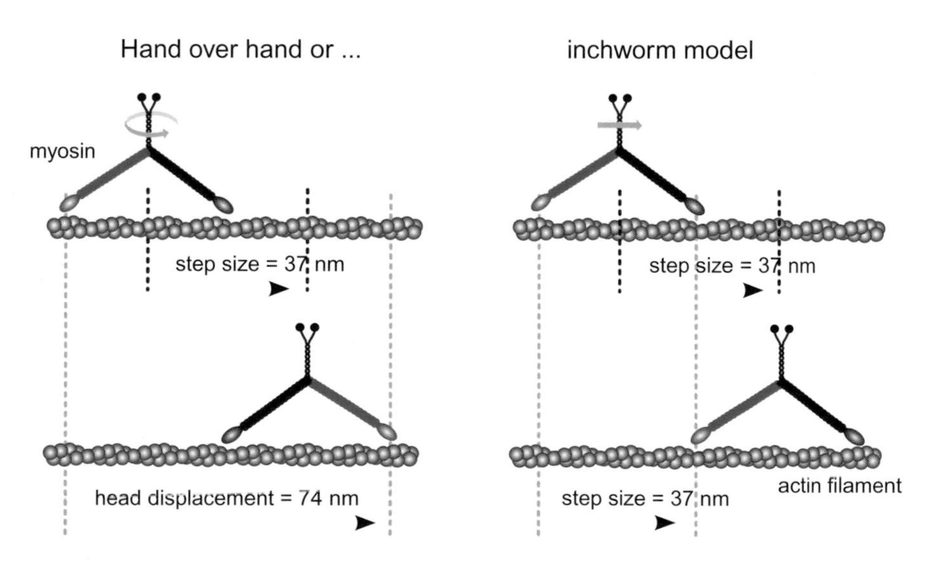

Figure 4.6. How myosin V moves along actin? Does it moves one foot ahead of the other each time or with always the same foot ahead of the other? In both cases the step size of the motor is the same (37 nm). But in the first case (a), at each step one foot of the myosin moves by 74 nm, while in the second case (b) the displacement is only 37 nm.

2.2 Myosin II: the motor of our muscles

Let's go back to myosin II, the main motor of our muscles, and also one of the most studied molecular motors. As we already mentionned this motor binds to actin with a single contact on actin. Its movement cannot be processive, so it cannot move on large distances without detaching from the actin filament. Consequently one can explain the movement of our muscles only by the activity of multiple motors working in parallel. This collective activity operates in a region of our muscles named sarcomere, where actin and myosins fibers are interconnected. In sarcomeres, myosins, when interacting with actin filaments, move along these fibers and produce a contraction (Fig. 4.7). A macroscopic movement is observable only if many myosins, linked together, can interact simultaneously with actin. When a myosin II detaches from actin after a movement a few nanometers long, it can bind again to the same filament because other myosins, still linked, maintain it in close proximity to the filament allowing further steps. Thus sarcomere's structure allows macroscopic movements but not the motor's lone one.

We can now wonder how a macroscopic force, large enough to allow us to move, is created from the force generated from a single myosin head. Micromanipulation techniques presented in Chapter 3 allowed the measurement of this small force. Yet experiments had to be designed specifically for myosin II since this motor is not processive. In order to observe multiple events, the experimental protocol consisted in maintaining the actin filament close to the studied myosin II. To do so researchers built a kind of dumbbell, in which two masses were handled with beads in optical tweezers with the actin filament stretched between them (Fig. 4.8). In the experiment a single myosin head, bound to a third bead, interacts with the filament. When the myosin is in contact with the actin filament, this filament is submitted to a traction. If the optical tweezers do not hold the beads too strongly the filament can almost move freely. One can then measure the individual step size, almost 5.5 nm in this case. If the optical tweezers hold firmly the actin filament, the filament motion is rapidly stalled and one can measure the force required to stall the motor from the very small distance travelled by the bead in the trap (see Chapter 3). The experimental result is a force of $\sim$5 pN. With the hypothesis that each step of the myosin II result from the hydrolysis of a single ATP molecule, one gets a yield of about 30% for this motor. All the energy is not used to produce movement, a large part of it is dissipated in the organism as heat, contributing to temperature elevation when practicing sport, which requires sweat to protect the body from overheating. From this few pN stalling force one can also deduce that more than 10^{12} myosin heads in parallel are needed to lift a 500 g weight, which

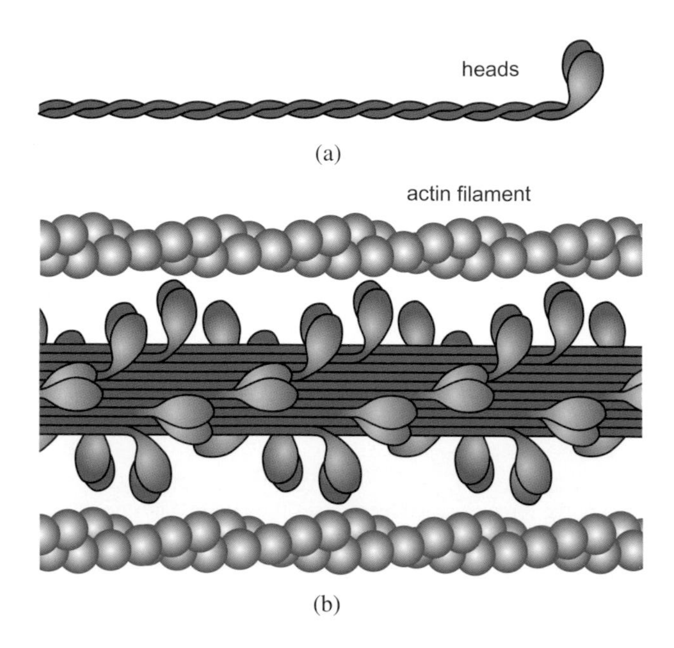

Figure 4.7. (a) In our muscles myosin II heads are associated pairs but the behavior of each head in independent of the other. (b) Muscle contraction results from the displacement of these numerous heads along actin filament. ATP hydrolysis provides the energy necessary for this movement.

demonstrates the need of many myosin heads acting together in the sarcomere to produce macroscopic movement.

Let's now investigate the link between chemistry and movement mechanics. To do so we must give a closer look at myosin II structure. Myosin II go by pairs, linked together with tails (Fig. 4.7), that allow them, at a larger scale, to form myosin filaments. Each single myosin has also a head, that can be bound to actin and whose structure is modified at different stages of its interaction with ATP. When ATP is present the head can detach from actin filament. Without ATP, the link between the myosin head and actin is very robust. This fact is at the origin of the "rigor mortis". When a organism dies, its F1-ATPases are no more active, ATP is no more produced, ATP already present in the body is degraded or consumed quickly, and then more and more myosin heads become linked irreversibly with actin and muscles become rigid. But what is happening in a living body? If one uses structural biology as a reading frame: with ATP myosin heads can detach from actin, structural modifications, deformations of a level arm, linked to ATP hydrolysis and ADP release occur and allow the motor to bind further on the filament, which creates a relative displacement of the myosin compared to the actin filament. This strong coupling between the structure modified by ATP chemistry and the related movement is sustained, as proven by many experiments. For example J. Spudich's

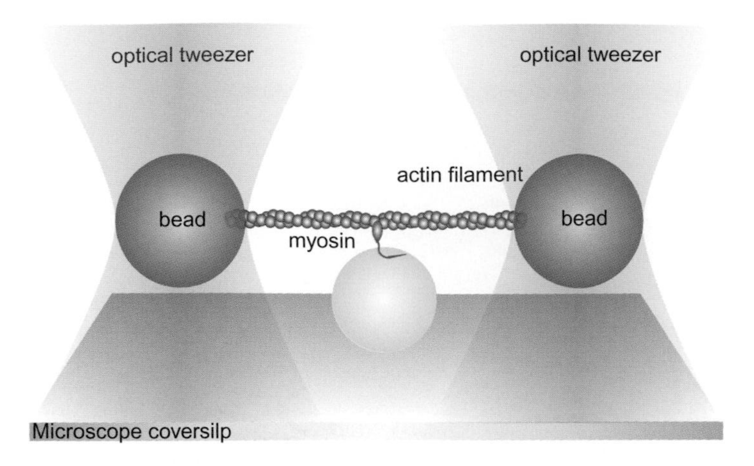

Figure 4.8. To measure the force excerted by myosin II on actin, one can use optical tweezers that hold firmly two beads attached to the extremities of a single actin filament.

group created mutants of myosin heads with longer level arms inducing longer step sizes, as the step size is proportional to the difference in length of the level arms. Yet a controversial experiment from T. Yanagida's group in Japan modified this simple description. This group could measure simultaneously ATP binding, thanks to fluorescent ATP, force and movement generation with optical tweezers. They observed that the movement could be delayed compared to ATP binding. Moreover, a single ATP binding could also generate multiple individual steps and even backward steps! These observations are not compatible with a simple link between structure and ATP chemistry. A general model, named thermal ratchet, was proposed. In this model motors are moving in a potential energy field that varies in time. Chemical reactions, that drive binding or unbinding to actin, are used to modify this energy profile, going from an asymmetric profile, as actin is an oriented polymer, to a potential, that, in a first approximation, can be considered as constant (see Fig. 4.9). When the potential is flat, myosin heads can diffuse symmetrically in both directions. But chemistry induces the asymmetric potential and myosin heads will then move towards the closest potential well. Kinetic parameters of the chemical reaction can be tuned to make the diffusion time, **on average**, larger than the time to diffuse over the nearest maximum, but smaller than the time to diffuse over the maximum in the opposite direction. From a physical point of view, asymmetry in a time varying potential and good kinetic parameters are sufficient to explain a directional movement and there is no need for a very strong coupling between chemistry and structure. In addition, as diffusion is a random process, fluctuations over an average movement can explain multiple steps as well as backward steps as observed by Yanagida's group with myosin (backward steps were also observed

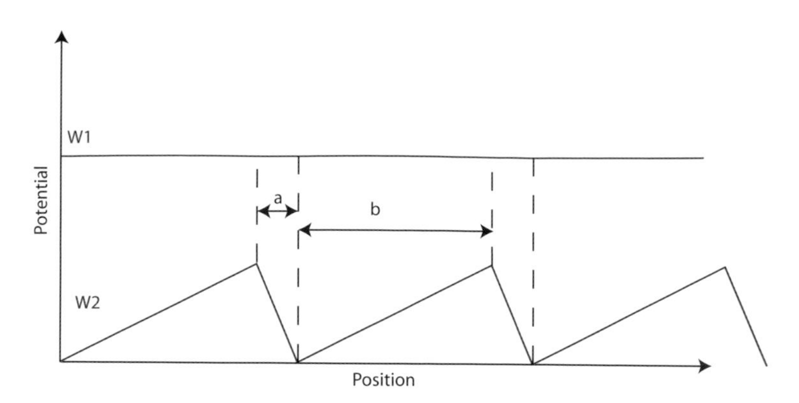

Figure 4.9. ATP binding and hydrolysis could modify myosin's interactions from an asymmetric potential energy profile (W2), when bound to the filament, to a relatively flat potential (W1) in which the motor can diffuse. In the asymmetric potential the motor will move to the nearest minimum of the potential. In the detached state (W1), motor can diffuse from its initial position (W2 minimum). If the time τ spent in (W1) state is such that the motor can diffuse over distances larger than a but smaller than b (so $a < \sqrt{D\tau} < b$, where D is the diffusion coefficient of the motor) then the motor will move **on average** towards the left. If chemistry is not properly adjusted ($\tau \gg \frac{b^2}{D}$), then there is no movement on average.

with F1-ATPase). For some motors there are still debates about which model is the best one to describe their activity.

To conclude on this part let's emphasize that understanding the behavior of a single motor is not always sufficient to understand the movement of a collection of motors. In some cases where motors are linked together, like myosin heads in our muscles, it is necessary to take into account the coupling between them to understand movement dynamics. Recent work has shown that this coupling could be at the origin of spontaneous oscillations in muscles. So there is still some work to fully understand how do an assembly of molecular motors like in our muscles really work.

3 A motor on DNA: the example of the RNA polymerase

Polymerases are presented in more details in Chapter 1. They are enzymes that use DNA as a template to chemically synthesize macromolecules, RNA for RNA polymerases and DNA for DNA polymerases, by using small molecules, called nucleotides as elementary bricks. During their movement along the DNA, polymerases use the chemical energy "freed" during the chemical synthesis to move along the DNA template. They use energy to move: they are motors! Contrary to myosin heads that do work on a rigid filament, actin, polymerases move along a very flexible polymer. Most of the time polymerase activity modifies elastic properties of the DNA substrate which provides new indirect ways to characterize their movement as will be shown on studies that involve RNA polymerase.

3.1 RNA polymerase and transcription

First recall some elements that are described in more details in Chapter 1, but required to understand the following part. As suggested by its name RNA polymerase synthesizes RNA, a macromolecule made of nucleotides that is essential in transcription, the process of going from the gene to the protein. To synthesize a specific RNA, RNA polymerase must read a DNA sequence from a starting point, named the promoter, to the end of the gene to be transcribed and must generate as few errors as possible. The first step for an RNA polymerase is to find and bind its promoter.

The second difficult step is to read the sequence. It is not so easy since, in DNA, the bases characterizing the sequence are not accessible since they lie inside the double helix. To perform this reading RNA polymerase will have to move into the double helix, modifying DNA structure by opening and unwinding the promoter region. Then the polymerase moves along DNA, using nucleotides freely diffusing in the surrounding medium, until it encounters the terminator region where it unbinds from DNA, freeing the RNA. Yet the real story is much more complex. An efficient method to study this mechanism uses labeled nucleotides to measure the size of the produced RNA molecules in order to deduce the movement of the RNA polymerase. But this method can give access only to values averaged over the whole population of RNA and cannot give access to subtle behaviors where unsynchronized fluctuations are important, nor to the role of mechanical constraints. Recent single molecule approaches have allowed, as we will now see, to have access to many details of RNA polymerase activity.

3.2 Setting the polymerase on the starting line

The very first step is the search mechanism of the polymerase, free in solution, towards its binding site, the promoter. It will not be addressed in this chapter as it will be described in Chapter 9.

Let's describe some experiments performed with magnetic tweezers in order to understand the first steps of polymerase's insertion into its promoter (Fig. 4.10). This insertion leads to local strand separation, that modifies local DNA topology and influences the whole DNA molecule. To understand these modifications consider a finite length DNA molecule. This double helix has a well defined number of turns, determined by the number of turns one strand wraps around the other. This number is a constant as long as DNA is not cut, or as long as no rotation is

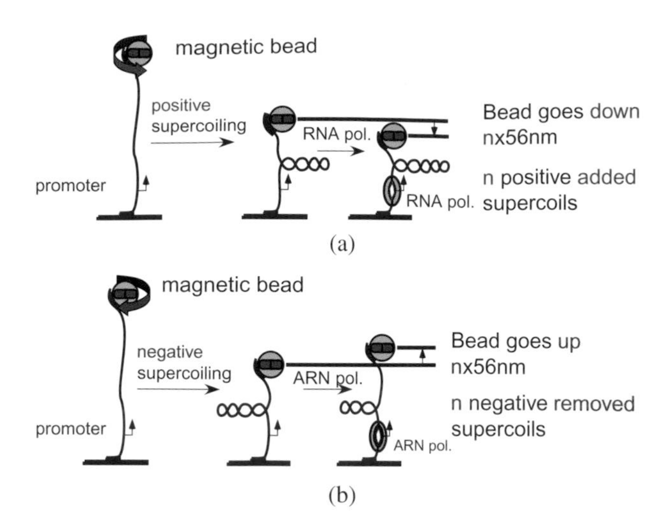

Figure 4.10. RNA polymerase binding induces strand separation at the promoter sequence. This melting generates a torsional constraint that produces supercoils or plectonemes. These plectonemes do modify the extension of the DNA molecule when it is placed under the mechanical constraint from stretched by magnetic tweezers. The different changes in the extension for different conditions i.e. positively (upper panel) or negatively (lower panel) allow us to measure the number of bases that are separated (inspired by Ref. [5]).

imposed to the molecule. When inserting into DNA, polymerase unwinds locally the DNA. As the total number of turns is constant it means that the rest of the DNA is overwound. By using micromanipulation techniques like magnetic tweezers (see Fig. 4.10 and Chapter 3) one can measure this unwinding. By adjusting the experimental parameters (in particular the force) the added turns in the molecule generated by RNA polymerase binding produce additional plectonemes that can be detected. But RNA polymerase binding itself also bends DNA and thus produces an additional shortening of the molecule. By studying RNA polymerase binding in different torsional conditions, T. Strick's group could differentiate these two effects and show that *E. coli* RNA polymerase opens ~13 bases and shortens DNA by 15 nm when it binds its promoter (Fig. 4.10).

3.3 A shillyshallying start

Once bound to the promoter RNA polymerase does not simply start its RNA processive synthesis. When analyzing the produced RNA one can observe very small fragments of only few bases (up to 10) that are synthesized before the final long RNA is produced. Thus RNA polymerase performs a series of synthesis that does not go to the end of the gene to be transcribed. Three mechanisms were proposed to explain these abortive cycles. RNA polymerase could go back and forth on the DNA

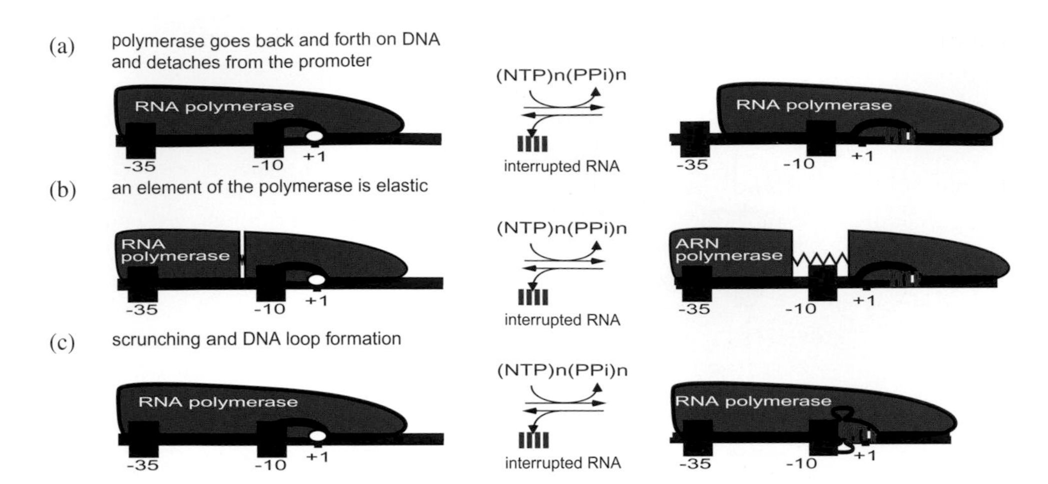

Figure 4.11. RNA polymerase is not only synthesizing full gene products, it produces also very small fragments. They are due to abortive cycles. To understand the molecular origin of these fragments T. Strick's group studied the distance variation between polymerase and its DNA substrate. Three models were initially proposed: a) The polymerase synthesizes but unbinds during synthesis after a few bases. b) The polymerase has an elastic part that is set under tension at the beginning of the synthesis. As long as the polymerase remains bound to its promoter, elastic tension makes it go back to its initial position. In some cases tension is such that polymerase can unbind from its promoter and perform a full transcription. c) The polymerase remains bound to its promoter and, by moving along DNA sequence, it pulls on DNA and forms a DNA loop. When the mechanical constraint is too strong, the systems goes back to its starting point. But sometimes this constraint allows the polymerase to unbind from its binding site and perform a full synthesis. This last scrunching mechanism was the one experimentally observed (figure inspired by Ref. [6]).

or it could stay bound to its promoter that would limit its unbinding probability, but consequently it would be unable to detach from the promoter and would then abort synthesis, resetting the polymerase to its starting position (Fig. 4.11). In this last scenario two schemes were proposed. In the first one the forefront part of the polymerase can move while the back end can only detach with a given probability, both being linked by an elastic part. It is only when the back end part of the enzyme detaches that the polymerase can perform the full synthesis of the gene. In the second model the polymerase does not move from its promoter but it can pull on the DNA "inside" the motor protein, forming a small DNA loop. In this case when the tension inside the loop is too strong, the forefront part detaches and only small transcripts are produced, but sometimes it is the back end part of the polymerase that detaches and then full transcription will occur. Two single molecule experiments, one using fluorescence and one using micromanipulations, investigated which scenario is happening in real life. S. Weiss' group utilized fluorophores placed on DNA and RNA polymerase and thanks to the FRET technique (see Chapter 2 and Ref. [7])

could measure the time variation of the distances between DNA and polymerase or between different parts of the polymerase during the synthesis. T. Strick's group used, like in the preceeding paragraph, magnetic tweezers to measure the global change in extension on the molecule, since if the scrunching mechanism occurs it should have a mechanical signature. Both approaches validated the scrunching mechanism. So at the initiation stage, the motor remains bound to the promoter and pulls on DNA to polymerize a few bases and comes back to the initial base. It performs some of these cycles until the motor, detached from the promoter, can escape and synthesizes the full RNA.

3.4 Some details about the RNA polymerase race

At every initiation cycle RNA polymerase may escape from its promoter and move processively along DNA: it can go into what is called the elongation phase that can last a few thousands bases. In this part we are going to focus on a few single molecule experiments developed to provide access to a detailed description of this elongation phase. In these experiments RNA polymerase is attached to a microbead manipulated with optical tweezers (see Fig. 4.12 and Chapter 3). This technique gives not only access to forces that can be exerted by the enzyme but also it allows us to measure polymerase movement with a subnanometric resolution.

Using this extreme precision, S. Block's group in Stanford could measure the elementary step size of the polymerase during its race. When working at a very low

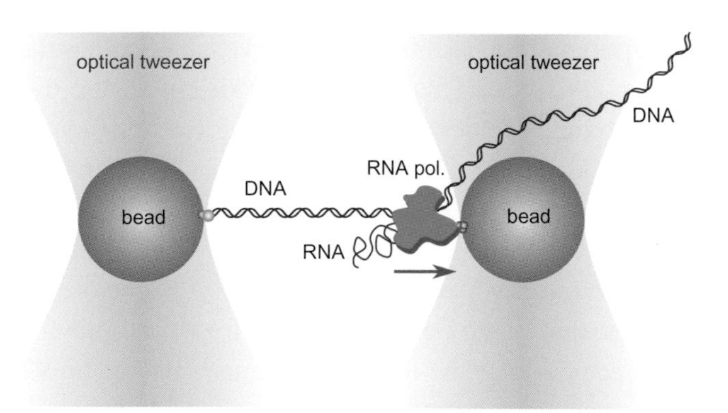

Figure 4.12. To study RNA polymerase during its race to transcribe, two optical tweezers are utilized: RNA polymerase is attached to beads held in optical tweezers and the extremity of a DNA molecule is likewise held in other tweezers. In this experiment RNA polymerase transcribes DNA is the direction of the force applied on the enzyme (green arrow) and a subnanometric measurement of the time evolution of the distance between the beads will provide all the information of the motor movement. It allows also the measurement of the elementary step size of the enzyme, its speed and the quantification of the pauses that may occur during transcription. (Adapted from [3].)

nucleotide concentration, in order to slow down the enzyme and to set conditions where these steps could be observable, researchers could measure a step of 0.37 nm, a distance very close to the distance between two DNA base pairs (0.34 nm). Independently a precise analysis of the relation between the RNA speed and the force applied on the RNA polymerase suggested that the motor follows the thermal ratchet model that we already introduced: RNA polymerase, at least the one from *E. coli*, seems to diffuse over a distance of about one base pair, the binding of a nucleotide blocks the enzyme in the next position.

RNA polymerase travel is far from being monotonous: it shows alternating periods with regular speed and pauses, but sometimes also backward steps. During the regular phase, RNA polymerase has an average speed of few nm/s (for *E. coli*) to few tens of nm/s for phage T7. This difference between polymerases from different species also exists, even though it is smaller between polymerases from the same origin. This "individual" behavior, not yet fully understood, could be observed only thanks to the high time and spatial resolution of single molecule studies using optical tweezers. Concerning the pauses during transcription, they occur every 100 bases on average and they can last up to a few seconds. Some antibiotics do increase the frequency of these pauses for some bacterial RNA polymerases in order to reduce the protein production rate and as a consequence, bacterial proliferation as well. Thanks to the extreme spatial resolution of their optical tweezers, S. Block's group could demonstrate that these pauses are correlated with the DNA substrate sequence and that their duration is very sensitive to the force applied on the DNA. A first consequence of this observation is that it is difficult to use the term stalling force for an RNA polymerase, as this stalling force is defined as the force required to stop the RNA polymerase. This is likely the origin of the discrepancy, from 14 to 25 pN, for the measurement of the stalling force between different groups. The second, and more important, consequence is that there are some biochemical steps inducing pauses that are sensitive to the force, so, these pauses must involve a variable conjugated to the force: it must be a displacement. Studies are now trying to relate these pauses to phases where the motor performs backward steps along DNA. Some longer pauses, that could be specific samples of the previous pauses, were also related to the error rate of the RNA polymerase. Indeed an RNA polymerase can, but rarely of course, incorporate a "wrong" nucleotide during the transcription, that is, it incorporates a nucleotide that does not correspond to the base read on the DNA template. In order to be as reliable as possible in the transcription process, RNA polymerase has a proofreading activity. It requires that after the detection of a misincorporated base, the RNA polymerase goes backwards with a single base pair step and cut the wrong nucleotide. To link these pauses and this proofreading mechanism the high temporal and spatial resolutions of single molecule studies will likely be essential.

3.5 From fundamental research to biotechnologies?

RNA polymerase pauses studies motivated S. Block's group to propose a DNA sequencing method based on these tools. Its principle is extremely simple. It relies on the ability to detect a pause with a single base resolution. During transcription, in order to move, RNA polymerase must bind the specific nucleotide complementary to the following base present on the DNA template. So when the solution around the polymerase is rich in base A, U, G but with a much lower concentration in C, the RNA polymerase pauses each time the DNA template has a G base where the RNA polymerase needs to bind a rare C nucleotide that has to diffuse to the catalytic site. As the position of these pauses can be determined very precisely they determine the position of the G bases along the sequence. To sequence the full gene transcribed the experimentalists just have to perform this experiment with solutions with low concentrations in A, then U, then G bases. This mechanical approach to sequencing is very elegant and simple in its principle but its large scale generalization seems quite unlikely. Other methods are now used in DNA sequencing and they allow us to sequence many sequences in parallel with relatively simple apparatus at quite a cheap price. In contrast this single molecule approach requires very sophisticated optical tweezers and cannot sequence many molecules in parallel, which is a required condition for DNA sequencing.

4 Conclusion

Single molecule micromanipulation techniques that were developed in the past decades have given a new perspective on molecular motors, and allowed the measurement of their speed, step size, yield or processivity (see Box 4.1). These physicist's tools could bring out conformational modifications related to chemical reactions, in which mechano-chemical coupling and Brownian fluctuations are important. Like it was in the case for RNA polymerase, or other enzymes studied with these tools, it is likely that similar experiments bring new important information on the many motors needed in living organisms.

Box 4.1. Summary for some molecular motors.

Motor	Role	Stalling Force	Speed (μm/s)	Step size	Processivity	Rendement %
F1-ATPase	ATP synthesis	Torque 40 pN · nm	~100 t/s	120°	Excellent	~100
Myosine II	Muscle contraction	5 pN		5 nm	Null	~20
RNA polymerase	RNA synthesis	30 pN	Variable typically 0.03	0.34 nm	Excellent	~20
Kinesine	Intra cellular movement	5 pN	1	8 nm	Good	~50
FtsK	DNA transport in bacteria during cell division	>60 pN	2	between 2.5 and 12 bases	Excellent	~50

Bibliography

[1] A general book on physics for biology: Phillips R, Kondev J, Theriot J. *Physical Biology of the Cell*, 1st Ed. Garland Publishing Inc (1 November 2008).

[2] Howard J. *Mechanics of Motor Proteins and the Cytoskeleton*, Sinauer Associates Inc., (2001).

[3] S. M. Block's group webpage to learn more about RNA polymerase and optical tweezers: http://www.stanford.edu/group/blocklab/

[4] K. Kinosita's webpage with videos and articles about F1-ATPase: http://www.k2.phys.waseda.ac.jp/

[5] Revyakin A, Ebright RH, Strick TR. Promoter unwinding and promoter clearance by RNA polymerase: Detection by single-molecule DNA nanomanipulation. *Proc. Natl. Acad. Sci. USA* (2004) **101**, (14) 4776–80.

[6] Revyakin A, Liu C, Ebright RH, Strick TR. Abortive initiation and productive initiation by RNA polymerase involve DNA scrunching. *Science* (2006) **314**, 1139–43.

[7] Kapanidis AN, Margeat E, Ho SO, Kortkhonjia E, Weiss S, Ebright RH. Initial transcription by RNA polymerase proceeds through a DNA-scrunching mechanism. *Science* (2006) **314**, 1144–7.

5

Cellular mechanics and motility

Sylvie Hénon, *Professor, Matière et Systèmes Complexes Laboratory, CNRS/Université Paris-Diderot.*
Cécile Sykes, *CNRS researcher, unité Physico-chimie Curie, CNRS/Institut Curie/Université Pierre et Marie Curie.*

The term motility defines the movement of a living organism. One widely known example is the motility of sperm cells, or the one of flagellar bacteria. The propulsive element of such organisms is a cilium (or flagellum) that beats. Although cells in our tissues do not have a flagellum in general, they are still able to move, as we will discover in this chapter. In fact, in both cases of movement, with or without a flagellum, cell motility is due to a dynamic re-arrangement of polymers inside the cell. Let us first have a closer look at the propulsion mechanism in the case of a flagellum or a cilium, which is the best known, but also the simplest, and which will help us to define the hydrodynamic general conditions of cell movement. A flagellum is sustained by cellular polymers arranged in semi-flexible bundles and flagellar beating generates cell displacement. These polymers or filaments are part of the cellular skeleton, or "cytoskeleton", which is, in this case, external to the cellular main body of the organism. In fact, bacteria move in a hydrodynamic regime in which viscosity dominates over inertia. The system is thus in a hydrodynamic regime of low Reynolds number (Box 5.1), which is nearly exclusively the case in all cell movements. Bacteria and their propulsion mode by flagella beating are our unicellular ancestors 3.5 billion years ago. Since then, we have evolved to form pluricellular organisms. However, to keep the ability of displacement, to heal our wounds for example, our

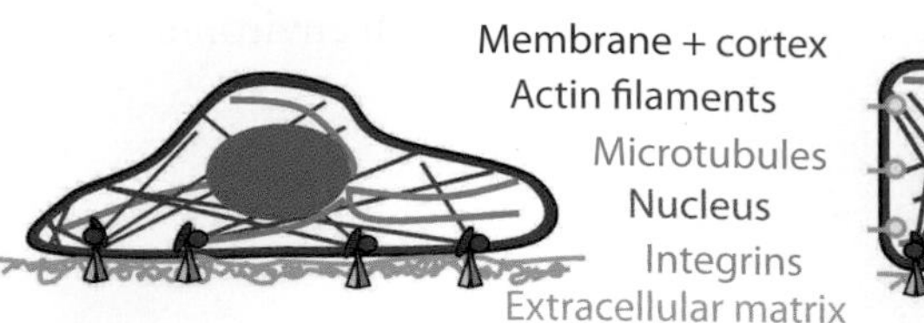

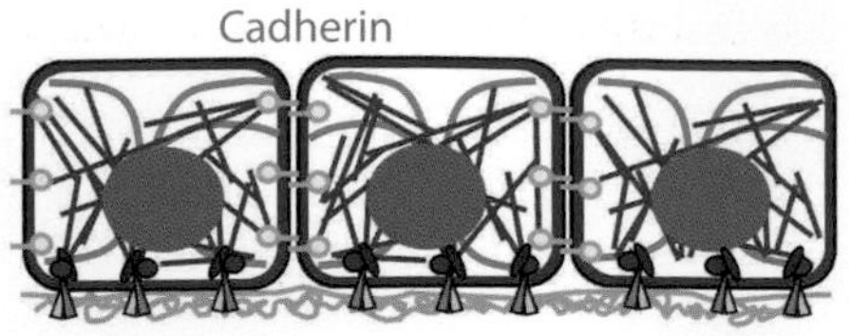

Figure 5.1. Simplified scheme of an isolated cell (left), and of a carpet of epithelial cells (right) with their cytoskeletons.

Box 5.1. Life at low Reynolds number [4].

Let us take a closer look at the hydrodynamic environment of bacteria of about 1 micrometer size. Bacteria moving in a fluid, like any other object, is subjected to two effects, one is the *viscosity* of the fluid, which tends to slow down the movement, and the other is the *inertia*, which allows movement once the object is started, even if it has stopped being powered. Let us translate each of these effects on to a swimmer in a pool: the swimmer launches into his lane taking a solid support on the wall of the pool, and can propel himself effortlessly over a few meters, *inertia* and its initial velocity allow him to move forward. However, his speed will decrease by the effect of the *viscosity* of the water in the pool. More specifically, the swimmer in the pool moves the fluid whose evolution equation (Stokes equation) reads:

$$\frac{\partial \vec{v}}{\partial t} + (\vec{v} \cdot \vec{\nabla}) \cdot \vec{v} - \frac{\eta}{\rho} \Delta \vec{v} = \text{propulsion term},$$

where t is time, $\vec{v}$ fluid velocity, η its viscosity and ρ its mass density. This equation has three terms on the left: the first describes the evolution of the velocity as a function of time and it is zero in the steady state, the second is the term of *inertia*, and the third term is for the *viscosity* effect. Let us propose a dimensional analysis of each of the terms of inertia and viscosity. For an object of size a that is moving at an average speed V in the fluid, the inertial term is $\frac{V^2}{a}$, while the viscosity term is $\frac{\eta}{\rho}\frac{V}{a^2}$. To compare the relative importance of the different terms, we define the Reynolds number R as the ratio of the inertial term and the viscous term. This number R is therefore $\frac{a\rho V}{\eta}$. Now, how much is R in a living system?

For a swimmer in a pool, R is of the order of 10,000. For a fish in a lake, R is of the order of 100, and for a sperm cell R is about 1/10,000. For objects of cell size, viscosity thus heavily dominates on inertia. To draw a more real picture, let us imagine what would be the hydrodynamic environment of a swimmer to be within the same hydrodynamic conditions as one of his sperm. The swimmer should be moving in molasses and swimming at a velocity of about 10 cm per hour. Ceasing to swim, he would stop in less than 1 μs. The life of a cell is thus at a low Reynolds number, and there is no inertia at the cell scale, any more than it is for our swimmer swimming in molasses and moving at less than 1 mm per second.

cells lost their flagellum, since it was not optimal in a dense cell environment: cells are too close to each other to leave enough space for the flagella to accomplish propulsion. The cytoskeleton thus developed inside the cell body to ensure cell shape changes and movement, and also mechanical strength within a tissue. The cytoskeleton of our cells, like the polymers or filaments that sustain the flagellum, is also composed of semi-flexible filaments arranged in bundles, and also in cross-linked or branched networks. It is a highly dynamical system in which filaments are able to elongate or slide one on the other with the contribution of very active cellular proteins like molecular motors. The versatile properties of this cytoskeleton ensure the diversity of mechanical behaviors to explain cell rigidity as well as cell motility.

1 Mechanical properties of eukaryotic cells

Cytoskeleton filaments organized into a network confer cells mechanical properties that we now describe. Those mechanical properties are a marker of "good health" of our cells that have thus the ability to resist mechanical stress or to adapt to external mechanical constraints. Cell mechanical properties are also characteristic of each particular organ: bone cells are far more rigid than brain cells. Moreover, cells are able to adapt their mechanical properties to a particular environment and we will describe recent advances in understanding how this mechano-transduction mechanism works.

1.1 Mechanical properties, rheology

Measuring mechanical properties of a cell or any material consists in measuring its deformation under an imposed mechanical stress. This is what is called rheology (Box 5.2). The first thing to do for studying cell mechanics is to design a cell rheometer, which means that we need a setup that allows us to apply a mechanical stress at the cell scale.

Box 5.2. Rheology measurements.

Rheology is the study of the visco-elastic properties of materials. When a body is subjected to a mechanical stress, it deforms, and visco-elastic properties are characterized by the relationship between the applied stress and the induced deformation. The deformation of the body is usually decomposed into two parts to account for the effect of stress: pure compression (or expansion), which corresponds to a volume change with a constant shape, and pure shear, which corresponds to a change in shape with a constant volume. This is illustrated Figure 5.1.

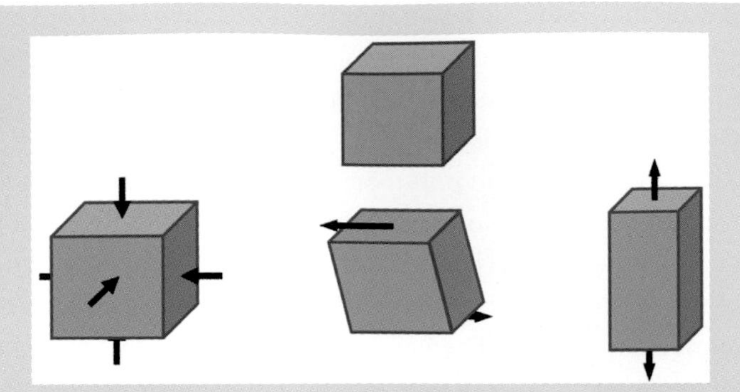

Three types of deformation of an object. Upper part: object before deformation. Bottom part: compression (left), pure shear (middle) and uniaxial stretch (right).

A stress is an applied force per unit surface. If the force is normal to the surface, the stress is a compression (or expansion) type of stress and equals the pressure. A shear stress corresponds generally to a force that is applied tangentially to the surface. For an incompressible material, the volume does not change, and therefore the material deforms by pure shear. A standard test in solid mechanics is a simple traction (uniaxial stretch), which is neither a pure compression nor a pure shear. It consists in pulling a bar on each end with a constant stress (Fig. 5.1 right). Its deformation is then measured through its strain ϵ which is a relative elongation $\Delta L/L$. For an elastic solid, the strain equals the ratio of the stress over the Young modulus E: $\epsilon = \sigma/E$. In order to completely characterise mechanical properties of a material, one can measure its response either to a constant step stress or to a sinusoidal stress. In the first case, one applies to the material an abrupt stress jump σ_0 that is kept constant, and the corresponding strain ϵ is measured as a function of time t. This leads to the creep function

$$Y(t) = \frac{\epsilon(t)}{\sigma_0}.$$

In the second case, the applied stress varies sinusoidally with time and the corresponding strain is studied as a function of the frequency f. It is characterised by two values: the ratio $G(f)$ of the amplitude of the stress over the one of the strain, and the phase shift $\varphi(f)$ between the strain and the stress. Alternatively, one single complex value is used $G^*(f) = G(f)e^{i\varphi(f)}$, and is called the complex modulus of the material.

$$\sigma(t) = \sigma_0 \cos(2\pi ft), \quad \epsilon(t) = \frac{\sigma_0}{G(f)} \cos(2\pi ft - \varphi).$$

For an elastic material, the strain follows the stress, the creep function is thus constant, and equals $1/E$ for a uniaxial stretch; the elastic modulus is also constant, and simply equals E, and $\varphi = 0$. On the contrary, a fluid flows as long as a stress is applied; for a Newtonian viscous liquid, the strain rate is proportional to the stress, and the proportionality factor is the viscosity η; in this case, the creep function varies linearly with time, $Y(t) = t/\eta$, and the complex modulus varies linearly with the frequency $G^*(f) = 2\pi fe^{i\pi/2}$.

Stresses and visco-elastic moduli are homogeneous to pressures, and the unit is pascal (Pa).

1.2 The cell skeleton or the cytoskeleton

The mechanics of cells depends on their cytoskeleton which was described in the first chapter. It is a network of filaments formed by self-assembly of proteins: actin, microtubules and intermediate filaments. Actin filaments and microtubules are polymers formed respectively of actin and tubulin monomers. Contrarily to chemical macromolecules, which are assembled covalently, cytoskeleton filaments are assembled mainly through electrostatic interactions. Within the cell, actin polymerization is possible thanks to a chemical energy source which is the hydrolysis of adenosin tri-phosphate (ATP) in adenosine di-phosphate (ADP).

Actin monomers polymerise at one end of an actin filament once in their ATP form (one actin monomer linked to one ATP molecule). Then, once assembled within the filament, actin monomers loose a phosphate, and subsequently depolymerize from the other filament end, and finally regenerate in their ATP-actin form within the cytoplasm. Actin filaments are thus asymmetric polymers that elongate mainly on one end, and depolymerize mainly from the other end. Microtubules elongate and shrink through an analogous process that consumes the hydrolysis of GTP (guanosine tri-phosphate) into GDP. These processes of elongation and shrinkage are responsible for a very high dynamics of the cytoskeleton. Moreover, molecular motors participate in the high dynamics since they are associated with cytoskeleton filaments and are able to slide them apart, or displace them along the cell membrane or the nucleus. The main players of cell mechanics and motility are actin filaments and their associated proteins. Some of them, like the cross-linking proteins, bridge actin filaments, either in parallel bundles, or imposing a fixed angle to the filaments, and allow the formation of actin fibers or gels.

A cooking analogy that can highlight the collective behavior of filaments and their mechanics is a spaghetti dish. If spaghetti are raw, they are rigid rods, and they do not confer any mechanical cohesive property to the spaghetti dish. If they are too cooked, spaghetti are very flexible and thus very curved in the dish; when one wants to take them out of the pan, they remain bulky, but can also slide between one another. Considering the scale of a cell, cytoskeleton filaments are rather reminiscent of *al dente* spaghetti: they behave neither like rigid rods, not like very flexible filaments, therefore, they are called semi-flexible filaments. They are thus able to entangle to produce an inseparable mass in the pan. Let us now imagine that each spaghetti is able to stick to a point of one of its neighbors, and that each spaghetti strand can stick to three different neighbors. One can easily imagine that, in these conditions, one single fork will be able to catch the whole dish, and the spaghetti lot will be able to deform when one pulls, in a certainly not reversible way. Even more complicated is the situation of the cytoskeleton where the equivalent of these sticky points can slide thanks to molecular motors able to walk on actin filaments or microtubules. This displacement relies on a conformation change of a part of the protein that consumes the energy of ATP hydrolysis.

The most common motors associated with actin are the myosin motors, and for example myosine II, which is the motor of muscular contraction. Myosin II is anyhow present in all cell types. One can already suspect that these molecular motors will be responsible for a certain viscosity of the actin gel (or network). Thus, the actin network containing myosin motors will appear less elastic in its mechanical behavior: a deformation will not be necessarily reversible because of this possibility of active sliding at the bridging points between filaments. The cytoskeleton forms in particular a shell of actin under the plasma membrane (the membrane that surrounds the cell), this shell is called the cortex. Its thickness is on the order of a few hundred nanometers and contains actin and myosin motors. Actin is equally present in other places inside the cell, and actin filaments can be as long as the cell size. Importantly, the actin cytoskeleton is also connected to transmembrane proteins, like integrins and cadherins (Fig. 5.1 right). This is what ensures the attachment of the cell to its environment, the extra-cellular matrix (proteic network in which cells are embedded), and the mechanical continuity of a tissue.

1.3 Cell rheometers

The characteristic size of a eukaryotic cell is small, on the order of tens of micrometers. In order to be able to apply controlled stresses, physicists and people like biomechanicians had to develop new rheometers. We will describe some of them. Microplates were used since 1930 to compress sea urchin eggs, and allow people to apply a uniaxial compression to one single cell, or to a cell aggregate. It is made of two parallel glass microplates (Fig. 5.2 left). It was originally used to measure cortical tension through shape analysis of the compressed cell or aggregate. Cortical tension σ is the energy per unit surface of the cell or aggregate, and is analogous to the surface tension of a liquid. The cell or aggregate shape at equilibrium is given by the Laplace law that links the difference in pressure across the membrane to the surface curvature

$$\Delta P = \frac{F}{S} = \sigma \left(\frac{1}{r} + \frac{1}{R} \right).$$

In a more modern version of the microplate rheometer [1], one of the microplates is rigid and the other one is flexible. The deflection of the latter allows measuring the force applied to the cell. This force can moreover be controlled through a feedback loop. An extension stress can also be applied with this setup, if cells adhere to

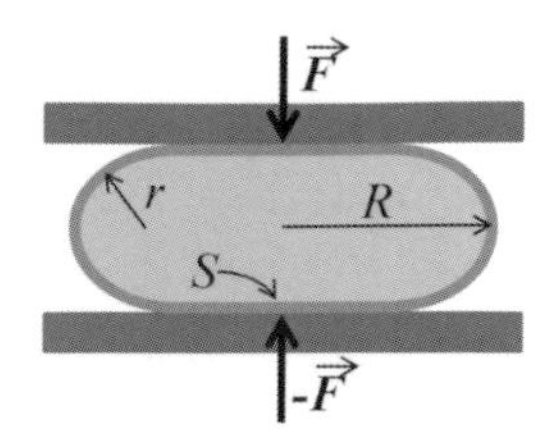

Figure 5.2. Microplate-based cell rheometer, left: in compression, right: in extension.

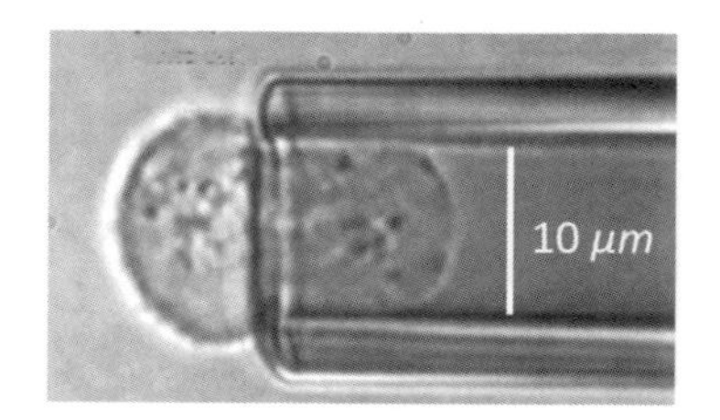

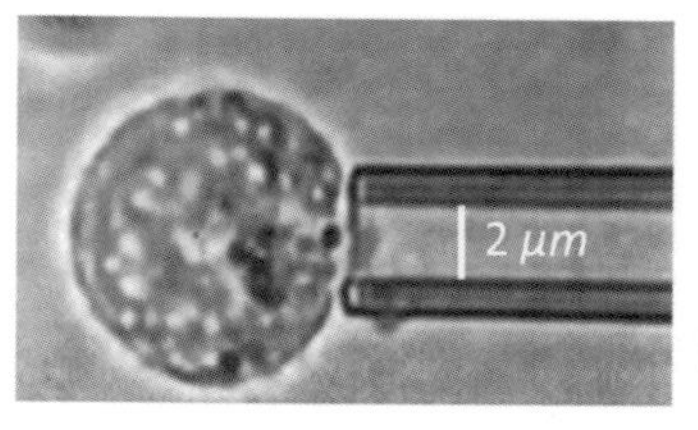

Figure 5.3. Micropipette-based cell rheometer, left: the cell is globally deformed, right, the cell is locally deformed.

the microplates (Fig. 5.2 right). With such a setup, the geometry is thus a uniaxial stretch, which allows obtaining the Young modulus of a cell, or the corresponding visco-elastic modulus.

Since 1950, micropipettes, with an inside diameter of 1 to 10 μm, have also been used to deform cells locally or globally by aspirating them (Fig. 5.3). The stress is here simply the aspiration pressure, but the deformation geometry is not simple. Analytical or numerical models thus have to be developed to link the deformation to a measurable parameter, like the length of the aspirated tongue.

The third family of cell rheometers uses beads as handles to apply stresses to cells, generally adhering to a substrate (Fig. 5.4 left). These beads are typically 1 to 5 μm in diameter. They can be inside the cell, but are more often outside the cell, and attached specifically to transmembrane proteins that link the outside to the cytoplasmic side of a cell (typically integrins or cadherins). Such beads can be trapped in optical tweezers used to apply a controlled force [1]. They can also be magnetic, and a magnetic field or a magnetic field gradient allows us to apply a controlled force momentum or force. The rotation or displacement of the beads can then be quantified when the force or the momentum is applied. In this case, the geometry is, again, not simple, and analytical numerical models need to be developed to understand rheological properties of cells. At last, atomic force

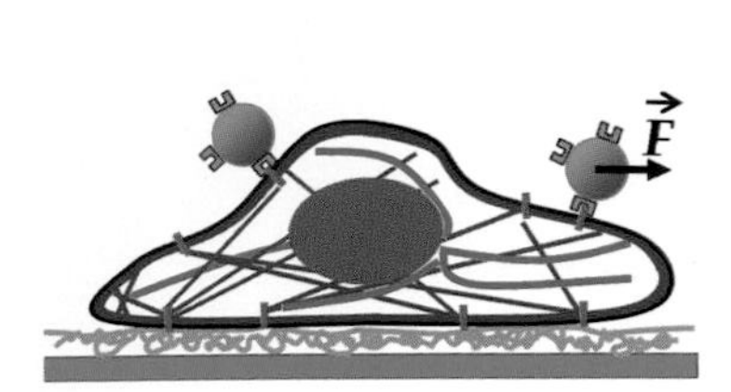

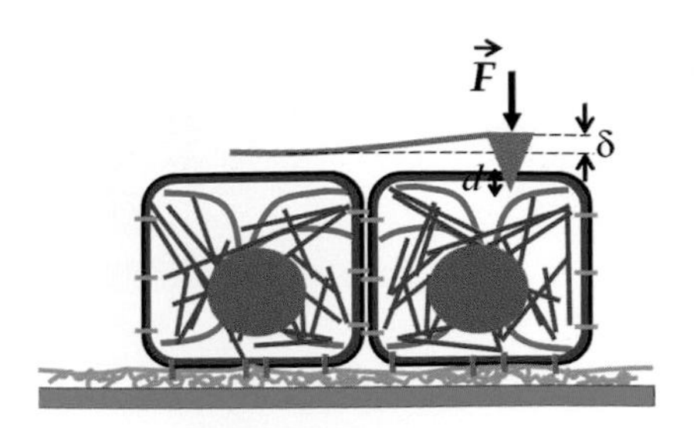

Figure 5.4. Cell rheometers for a local measurement: left, bead rheometer; right, atomic force microscope (AFM).

microscopes (AFM) are used to quantify cell mechanics at a very local scale, that of an AFM tip, from a few nm to a few μm (Fig. 5.4 right).

Cell rheology can also be made otherwise, not by applying mechanical stresses, but by analysing the spontaneous movement of small objects: this is called passive microrheology. Small particles, of diameter typically below 1 μm, can be naturally present in cells (for example granules), or microbeads can be injected, or can adhere to the cell surface. Such beads undergo continuous movement under fluctuating forces. In a passive material, the origin of fluctuating forces is the thermal agitation of molecules, which is well known as Brownian motion. The amplitude of the Brownian movement of beads is directly linked to the visco-elastic modulus of the material through temperature by a generalization of the famous Stokes-Einstein formula.

For a living cell, things are more complicated, since forces exerted by molecular motors add to the thermal agitation within the cell. Passive microrheology can thus be used either to characterize cell visco-elasticity if fluctuating forces are well known (for example keeping the high frequency part where thermal agitation dominates), or, on the contrary, to characterize forces exerted by molecular motors. Through all the above described methods, rheological properties of cells can be quantified, on a wide range of characteristic sizes and times. Such properties are not straightforward to understand, at short or long times, as we will now address.

1.4 Rheological properties of cells at short times

It is important to note that eukaryotic cell volume is in general, constant: cells are incompressible, they deform at constant volume, in pure shear. Under a constant stress applied on short times (from a few tens of milliseconds to a few minutes), cells continuously deform (they creep), but more slowly than simple liquids: their strain ϵ does not linearly increase as a function of time t, like for a liquid, but varies as a weak power of time: $\epsilon(t) = \beta t^{\alpha}$, with α between 0.15 and 0.35 (Fig. 5.5 left).

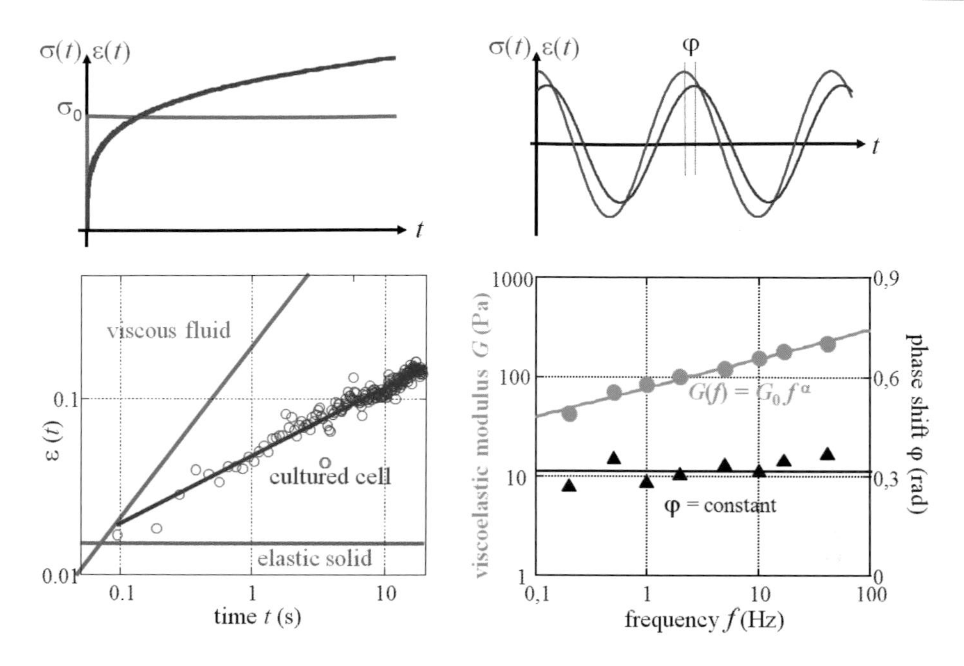

Figure 5.5. Creep (left) and complex modulus (right) measurements on isolated cells.

Similarly, at "average" frequencies (from a few tens of mHz to a few tens of Hz) the visco-elastic modulus varies as a power of the frequency, with the same exponent α; on the other hand, the relative phase shift of the strain over the stress is independent of the frequency: i.e. $G^*(f) = G_0 f^\alpha e^{i\varphi}$ with φ independent of the frequency (Fig. 5.5 left).

The characteristic value of the visco-elastic modulus G thus depends on the time or frequency range over which it is measured. Depending on the cell type, its value spans from 100 Pa to 10 kPa for a characteristic measure time of 1 s, while the Young modulus of rubber is typically 1 MPa and the one of diamond is 1,000 GPa. Cells are therefore very soft objects. There is a high variability in G values within a population of a single type cells. On another hand, the average value of G is well characterized in one cell type. However, cells are heterogeneous objects: G ranges from typically a few kPa for the nucleus, to a few Pa or tens of Pa for the cytoplasm, and the cortex is characterized by a few kPa to a few tens of kPa. As a summary, cells creep proportionally with a power of time, having a mechanical behavior that is intermediate between the one of a solid and the one of a liquid.

Other materials are known for their power law rheology. First, heterogeneous media out of equilibrium like concentrated emulsions, mud, foams, that are considered as soft glassy materials; then gels with a fractal structure. It is tempting to draw an analogy between cells and each type of these materials. The same rheological law as a power function of time, in cells, measured at all length scale, from a few nm with an AFM to 10 μm with uniaxial stretch, is reminiscent of a multi-scale structure, which is consistent with the fractal structure of the cytoskeleton. Moreover, cells are also heterogeneous, out of equilibrium media, and thus could be analogous to soft glassy materials. No clear argument allows us to distinguish for the moment between one model or the other.

1.5 Forces generated by cells

Cells are not mechanically inert: they are able to exert forces on their environment. Therefore, when a cell adheres on a thin elastic film, folds form on the elastic film because of tractive forces exerted by the cell (cf. [2]). Those forces can be measured through the deformations they induce on a deformable substrate. This substrate can be a hydrogel in which fluorescent micron-sized beads are included (cf. for instance [3]): bead displacement will allow us to derive the field of forces generated by cells. Micropillar substrates can also be used, they are made of elastomer (Fig. 5.6). Under forces exerted by cells, these micropillars are deflected, and their deflection is a measure of the applied force [1].

At last, the force generated by a whole cell can be measured by the microplate technique (Fig. 5.2): a cell that is spread between two microplates exerts a tractive force that can be measured through the deflection of the soft microplate. All these experimental setups lead us to evidence that an adhering cell exerts tractive forces on its surrounding. The highest forces are generated on the edge of isolated cells, or at the periphery of cell aggregates, where tractive forces are on average 2 to 4 times higher than in the middle of the cell or the aggregate. Moreover, the cell has the ability to adapt to its mechanical environment: forces generated by cells are greater as the substrate is stiff. In the case of micropillar measurements, the average force, as well as the maximal force, exerted on a single micropillar of typical diameter 1 μm is directly proportional to the micropillar stiffness and can span by two orders

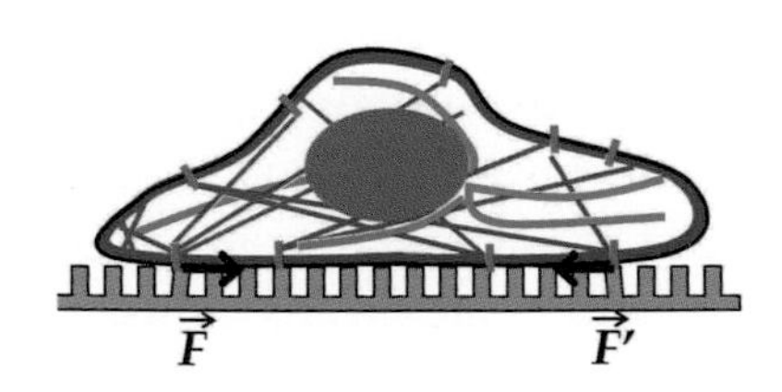

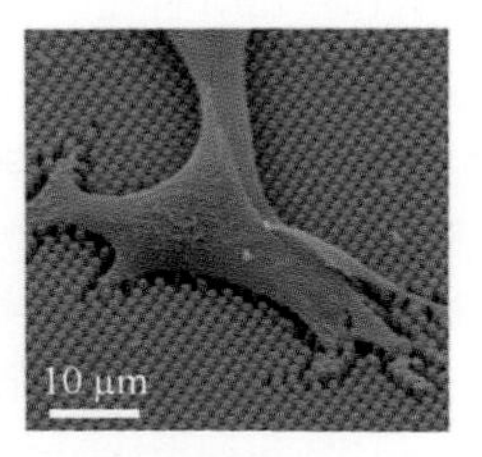

Figure 5.6. Elastomer micropillars with an adhesive tip for cells allow us to measure forces exerted by cells. Left: scheme. Right: Electron microscopy images of cells adhering on a carpet of micropillars.

of magnitude; however, above a threshold, forces saturate: the maximal force a cell can apply locally on a flat substrate is about 10 to 20 nanonewtons.

Similarly, cells adhering on microplates generate a maximal tractive force that is proportional to the microplate stiffness, until it saturates at a few hundreds of nanonewtons. Molecular motors are generally responsible for generation of forces in cells. Molecular motors belong to a class of proteins able of exerting forces to walk along actin filaments or microtubules. In particular, the action of molecular motors put the cytoskeleton under pre-stress. Therefore, the cell cortex is under tension (like a stretched spring), and cultured cells kept on a substrate develop tension fibers, long actin and myosin structures that cross the cell from one attaching point to another.

1.6 Rheological properties of cells at long times, and mechanotransduction

When applying mechanical stress to cells over long periods (typically beyond several minutes), they show an active response: cells adapt to mechanical stress applied to them. This is called mechanotransduction. For example, as a result of pressure or repeated friction, the skin cells of the fingers of a guitarist proliferate and calluses develop. At the scale of a single cell in culture, some cells increase their elastic modulus by an order of magnitude in a few minutes when they are under prolonged or repeated stress. More generally, cells are sensitive to mechanical properties of their environment and adapt to it.

For instance, adherent cells grown on hard substrates (the plastic cell culture dishes, glass) will proliferate, while grown on too soft substrates, they will go into apoptosis (cell death). This could play a role in the proliferation of cancerous tumors. This raises the question of elucidating the mechanisms of mechanotransduction. We know that the points where a cell adheres to the extracellular matrix are the ones that act as mechanical sensors. More specifically, some adhesion proteins can be stretched while under the influence of a force. This stretching reveals a cryptic site, which is hidden when the protein is at rest, but becomes accessible when stretched. A biochemical reaction will begin by protein binding to this site, triggering a cascade activation of actin polymerization, which will mechanically reinforce the cell. Cell therapy has recently been confronted with the sensitivity of cells to their mechanical environment. Progress on the isolation and culture of stem cells can allow considerable regeneration of damaged tissues from stem cells.

But it has been realized that it was not enough to introduce stem cells in a destroyed tissue to see it rebuild. For example, if stem cells are injected to a patient who had a myocardial infarction, very few of these cells become heart cells, because the necrotic zone is very rigid. We also know that after a muscle tear, complete immobilization leads to calcification rather than muscle regeneration: to regenerate muscle fibers, myoblasts, premuscular cells, must be subjected to mechanical stress. With these lessons, research in tissue engineering has developed reactors where the geometric, biochemical and mechanical environment of stem cells are tightly controlled to create a new tissue.

2 Cell movement or cell motility

We described in the introduction flagellar movement, which is for example how our sperm cells move. This allowed us to easily define the cellular hydrodynamic environment (See box on life at low Reynolds number). However the cells of our tissues do not move by flagella, but through an incessant reorganization of their internal cytoskeleton. The same hydrodynamic conditions are of course valid for the cells of our tissues that move in an environment dominated by viscosity.

Cell movement of mammalian cells in general can be described by a simplified three-step process: the first step is the extension of the membrane at the front, the second step is the cell adhesion to the substrate on which it moves (it crawls), and finally the third step is the retraction of the cell body (Fig. 5.7). A cell in contact with a substrate follows the substrate shape, whether plane (the bottom of a Petri dish) or like a three-dimensional extracellular fiber network, a more physiological situation. At the front of the cell in the direction of motion, the membrane forms a fine veil parallel to the substrate, and this veil is underlined by the cables of the actin cytoskeleton, which elongates by polymerization and pushes forward. The cell develops transitory adhesive points with the substrate, and the rest of the cell body follows the net movement by retracting under the effect of molecular motors that slide along actin filaments.

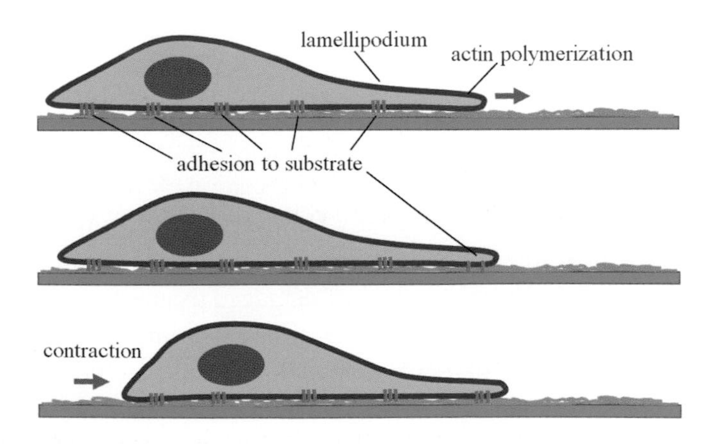

Figure 5.7. The cell moves over a substrate by a three-step mechanism, as described in the text.

2.1 Elongation of the lamellipodium

We have described in "cell skeleton" or "cytoskeleton" how actin filaments polymerize at one end and depolymerize at the other. This mechanism allows for a continuous actin movement that does not involve the translocation of actin filaments, like a kind of track movement of the tank, the track being the actin filament, generally immobile, but growing on one side and shrinking on the other. At the front of the cell, in the direction of movement, the elongation of the actin filaments by polymerization deforms the fold of the membrane (or "lamellipodium"). Therefore, the actin filaments extend just below the membrane, by activation (the equivalent in the macromolecular chemistry would be "catalysis") of their polymerization through specific activator proteins. The activation of actin polymerization at the cell membrane allows the filaments to push forward the membrane. Note that the force generated by actin polymerization is sufficient to deform the membrane.

Let us look at its magnitude. First, note that the energy required to bend a membrane is on the order of a few k_BT; in other words the membrane fluctuates under the influence of thermal fluctuations. Indeed, a liposome membrane which is not tense appear under the microscope as fluctuating as a function of time. An actin filament has a thickness e of five nanometers, therefore the force generated by polymerization, necessary to deform the membrane, should be on the order of k_BT/e. Thus the unit of measure is the piconewton: this is an estimate of the scale of the force required for an actin filament to deform the membrane of the cell. In fact, the pushing force that generates motion by actin polymerization is not only due to the action of parallel filaments, but also due to the branched and intertwined, even cross-linked, filaments. This is thus a real gel (within the meaning of rubber gels) that pushes the membrane of the lamellipodium.

2.2 Cell adhesion to its substrate

For the crawling movement of the cell to be effective, there must be cell adhesion to the substrate on which it crawls. Adhesion points are thus created, which stabilize the cell on its substrate. These points of adhesion have a characteristic life time of a few minutes and can come off more easily as a result of the contraction of the cell body: they "mature", and detach at the rear of the cell to allow the cell to move forward.

2.3 Cell body retraction

The third step of cell movement is the retraction of the back of the cell, by contraction of the shell of the cytoskeleton which is just underneath the membrane (the cell cortex). This contraction, very similar to that of a muscle, is generated by the sliding of myosin molecular motors on the network of cortical actin filaments.

2.4 Hijacking the cell machinery

The cell motility machinery is also hijacked by intracellular bacteria or viruses for their own movement. For example, bacteria such as *Listeria monocytogenes*, which cause the desease of listeriosis, move within a cell by actin polymerization. This polymerization takes place at the surface of the parasite with an activator that triggers the local growth of actin filaments using proteins from the host cell. This same mechanism of actin polymerization is, as we described above, also used by the lamellipodium for its extension.

About a hundred of filaments work together to form a "comet" that propels the parasites. In fact, bacteria like Listeria consitute a simplified living system model to study the general mechanism of force and movement generated by actin polymerization. For example, these bacteria can be isolated by centrifugation, and it is thus possible, by characterizing the proteins present at the surface of the bacterium, to identify the proteins responsible for actin assembly. Thus, it is possible to reconstruct the movement of actin polymerization in the presence of isolated and purified cellular proteins.

3 Simplified systems for a controlled study

The living cell is an extremely complex system. To unravel this complexity, we can make a detailed description, and then seek to deconstruct the essential components, specifically inhibiting certain molecules or certain interactions between these molecules. Thus, for example, the central role of actin and myosin in mechanics and cell contractility can be highlighted: when actin filaments are depolymerized, the rigidity of the cell falls; when myosin is inhibited, the cell loses its contractility. But this approach is not sufficient. Indeed, a complex system is composed of interconnected parts and has properties that are not simply the sum of the properties of its individual components. In a complementary approach, we can instead seek to reproduce biological behavior from a limited number of components and thereby capture the minimum ingredients to reproduce certain cell functions like shape changes and movement.

3.1 Actin gel rheology

The visco-elastic properties of actin filament solutions have been studied in detail with the goal to understand the cellular rheology. *In vitro* concentrated solutions of actin filaments can be obtained from purified proteins. Adding cross-linking protein, such as filamin A, gels display a power-law rheology, like living cells. However, gels of purified actin are much softer than cells, even at high actin concentrations

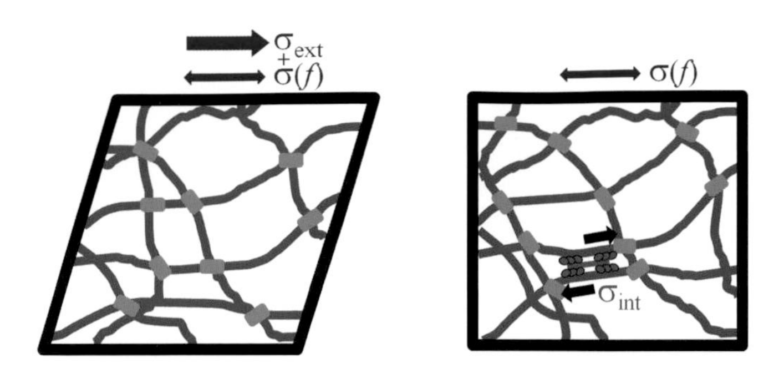

Figure 5.8. Measuring the rheological properties of actin gels containing protein crosslinkers. Left: Measure of the response to an oscillating stress superimposed to a static preload. Right: Measure of the response to an oscillating stress of a gel in which myosin complexes apply a pre-stress.

corresponding to physiological conditions: the characteristic viscoelastic modulus, at about 1 Hz, is only a few Pa. In the case of pre-stressed gels, the elastic modulus is 1,000 times higher and is consistent of the rheological behavior of the cell cortex. This pre-stress can be understood *in vivo* as the effect of the action of myosin molecular motors assemblies, which put actin filaments under tension. However, a purely mechanical pre-stress, obtained by shearing the actin gel, has the same effect (Fig. 5.8). Therefore, actin filaments, cross-linking proteins and pre-mechanical stress are the three ingredients needed to reproduce the mechanical behavior of the cell cortex.

3.2 Symmetry breaking of a gel grown around a bead: mechanics and dynamics

To better understand the properties of actin gels growing from a surface, such as the lamellipodium, simplified experimental systems inspired by the bacterium Listeria, which hijacks the cell machinery, are an alternative means to study cellular complexity. These systems are made of rigid or deformable beads, the surface of which is covered by activators of actin polymerization. The gels grown from these surfaces have the same properties as the gels described above: they are visco-elastic. However, let us first illustrate their elastic properties.

Let us take a look at the experiment of Fig. 5.9. Actin is fluorescent and appears bright. The actin gel grows from the bead surface on which actin monomers assemble, and is visualized. A cohesive gel thus grows in a spherical geometry, centripetally from the surface and therefore undergoes a mechanical constraint. As a consequence, a stress accumulates gradually as the actin gel grows. The shear stress is greater at the outer edge of the gel than at bead surface (where it is zero), while a normal stress accumulates on the surface of the bead. The high shear stress at

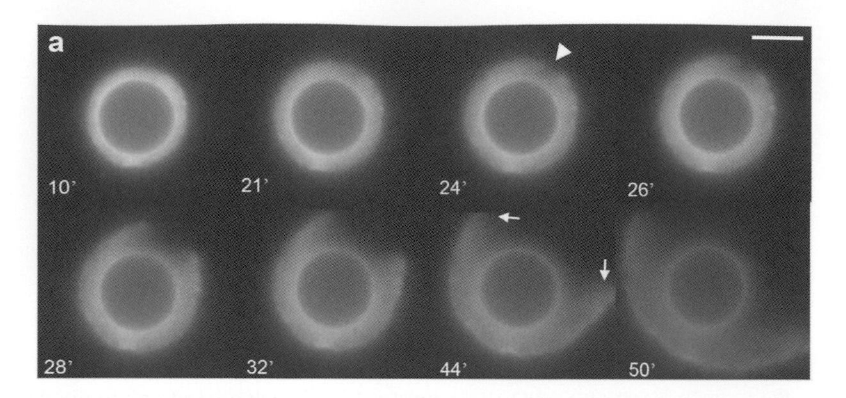

Figure 5.9. Sequence of images of the symmetry breaking event of an actin gel grown from the spherical surface of a bead. Time is indicated in minutes. Bar, 10 micrometers. Figure reprinted from Ref. [5] (Copyright 2005 National Academy of Sciences, USA).

the outer surface of the gel may cause the breakage of the gel as soon as the stress exceeds a certain threshold. This is what is seen in Fig. 5.9, where the point of breakage is indicated by a white arrowhead. One can notice that the stress relaxes after breakage, since at 50 minutes the outer layer appears wide open. The propulsive comet is subsequently developed on the opposite side of the point of symmetry breaking.

3.3 Movement of inert objects on the living system model

Once the symmetry breaking is completed, a polystyrene bead is able to move under the effect of the continuous polymerization of actin. But what is the mechanism of this movement? What is the role of mechanics on this movement? To highlight the effect of actin gel mechanics during movement, oil droplets can be coated by an activator of polymerization and placed in conditions of actin polymerization (Fig. 5.10).

Note that the droplet is deformed on the side where the comet is, at the back of the movement. The effect of the actin gel growing from the surface of the droplet is to compress the droplet. The normal stress exerted by the actin gel on the droplet can be quantified by using the Laplace equation, which expresses that the pressure difference between the inside and outside of an interface is proportional to the surface tension and inversely proportional to the radius of curvature of the interface. This leads, for a spherical surface, to $\Delta P = 2\gamma/R$, where ΔP is the difference in pressure between the inside and the outside, γ the surface tension, and R the curvature radius of the interface. With R being the curvature radius of the droplet's right side (undeformed), and r the curvature radius of the droplet just underneath the comet, on the left side of the droplet the Laplace equation can be written as

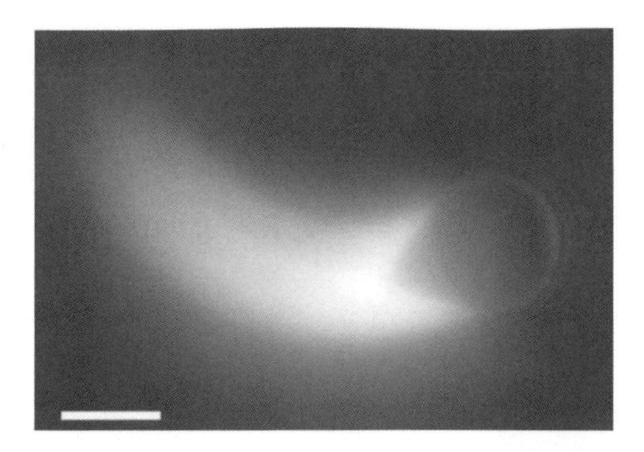

Figure 5.10. An oil droplet is covered by an activator of actin polymerization and placed in a cell medium containing fluorescent actin, which assembles into a comet. Bar, 4 micrometers. (Figure taken from [6].)

$\Delta P - \sigma = 2\gamma/r$, where σ is the normal stress exerted on the droplet by the growth of an actin gel, which is zero on the right side of the droplet.

An estimate of σ is $30\,\mathrm{nN}/\mu\mathrm{m}^2$ in the biochemical conditions of the experiment of Fig. 5.10. Note that the value of σ depends on the elastic properties of the actin gel, in particular its elastic modulus, and the stress will be much greater when the elastic modulus is higher. Thus actin-based movement by polymerization generates compressive forces able to deform an interface. The bacteria *Listeria*, as much as activated beads, move therefore through this effect of rear compression, reminiscent of a pinching of a cherry core between the thumb and forefinger to make it slip out of grip and roll on the ground over to the other end of the garden. However here in the case of actin-based motility, there is no inertia; unlike in the case of cherry core, neither the bead nor the *Listeria* will continue to move after the actin polymerization has been stopped.

4 Conclusion and perspectives

Our cells, and mammalian cells in general, have complex visco-elastic properties, that allow them to react to the constraints imposed by the outside, either through the extracellular matrix or through the family of cells of the same tissue type to which they belong. Our cells have the ability to move within our organism, for our good (wound healing), but also for our pain (cancer metastasis). Cell rheology and motility can be understood through experiments on single cell or *in vitro* reconstituted systems that are analyzed using concepts coming from the physics of polymers or macromolecules. Therefore, the study of cell rheology is at the crossroad junction between different disciplines such as biochemistry, mechanics, physics.

Moreover, cellular systems bring new issues that are not addressed elaborately enough in physics or mechanics, such as the understanding of cellular activity in response to mechanical stress, a phenomenon that does not exist in physical chemistry of macromolecules. We focused here on the mechanical properties and motility of lone single cells, however, the cell should be treated as if it were in its physiological environment, like in the extracellular matrix and in the presence of neighboring cells. Some pathological cells have been shown for example to be softer than the same cells in their normal state, or to migrate faster once isolated. However, the tissue they form may be stiffer than the normal tissue, and the cells may also migrate differently. This kind of effect can be explained either by a strong adhesion between cells, either by an increase of cellular adhesion, or by the reaction of the cell to its environment such as an increased production of extracellular matrix cell, or a combination of these causes.

Finally, the understanding of cell mechanical properties allows to consider the design of synthetic materials with mechanical properties suitable for controlled cell growth, paving the way for new paths in the reconstruction of damaged tissues. These areas of physics and biology are therefore also having interdisciplinary overlaps with the field of medicine, making these approaches very rich and complex subjects for future research.

Bibliography

[1] Asnacios A, Balland M, Biais N, Cardoso O, Desprat N, Du Roure O, Gallet F, Gerbal F, Hénon S, Ladoux B, Richert A, Saez A, Siméon J. Des outils de micromécanique pour comprendre l'adhésion et la migration cellulaire. *Bull. Soc. Fr. Phys.* (2005) **148**, 4–8.

[2] Harris AK, Wild P, Stopak D. Silicone rubber substrata: a new wrinkle in the study of cell locomotion. *Science* (980) **208**, 177–9.

[3] http://www.cellmigration.org/resource/imaging/imaging_approaches_force_imaging.shtml

[4] Purcell EM. Life at low Reynolds number. *Am. J. Phys.* (1977) **45**, 3–11.

[5] Jasper van der Guch et al., Stress release drives symmetry breaking for actin-based movement. *Proc. Natl, Acad. Sci.* **102** 7847–52.

[6] Boukellal H, Campàs O, Joanny J-F, Prost J, Sykes C. Soft listeria: actin-based propulsion of liquid drops. *Phys. Rev. E* (2004) **69**, 061906.

6

Exploring neuronal activity with photons

Laurent Bourdieu, *CNRS researcher, IBENS, École normale supérieure, Paris.*
Jean-François Léger, *CNRS researcher, IBENS, École normale supérieure, Paris.*

1 Introduction

In the nervous system, sensory information and motor behaviors are encoded by dynamical activity patterns of neural networks. Sensory information collected by the sensory organs is transmitted by a series of neural networks from the periphery of the nervous system to the central nervous system (Fig. 6.1). Within it, a hierarchy of brain areas performs more and more complex calculations. Different data collected in the same modality (the position and color of an object detected visually) or from different modalities (the position of an object measured by vision and audition) are assembled (Fig. 6.1). At the end of this sensory integration, specific activity patterns are ultimately generated in populations of neurons within motor areas, so as to produce the appropriate behavior in relation to the sensory information (escape, attack, move, etc) (Fig. 6.1). In this cascade of events, a key step of information processing seems to occur at the level of local microcircuits containing a few thousand cells. Sensory cortical areas are organized in repeating units called columns. These columns, perpendicular to the cortical surface, have a section of about $0.5\,\text{mm}^2$, contain about 10^4 neurons and can be considered as a fundamental component of the calculation. All these microcircuits are highly-connected three-dimensional networks composed of projection neurons (inputs and outputs of the local loops) and local inhibitory and excitatory neurons. In addition, these networks form a dense network with surrounding glial cells. Despite the fundamental role of these microcircuits in information processing, their operating principles remain largely to be discovered. How local networks of neurons process incoming signals and integrate them in the presence of permanent brain functional activity is indeed still poorly understood. The main reason is that the traditional experimental tools based on electrophysiology cannot record *in vivo* simultaneously and comprehensively the activity of all neurons forming such a microcircuit.

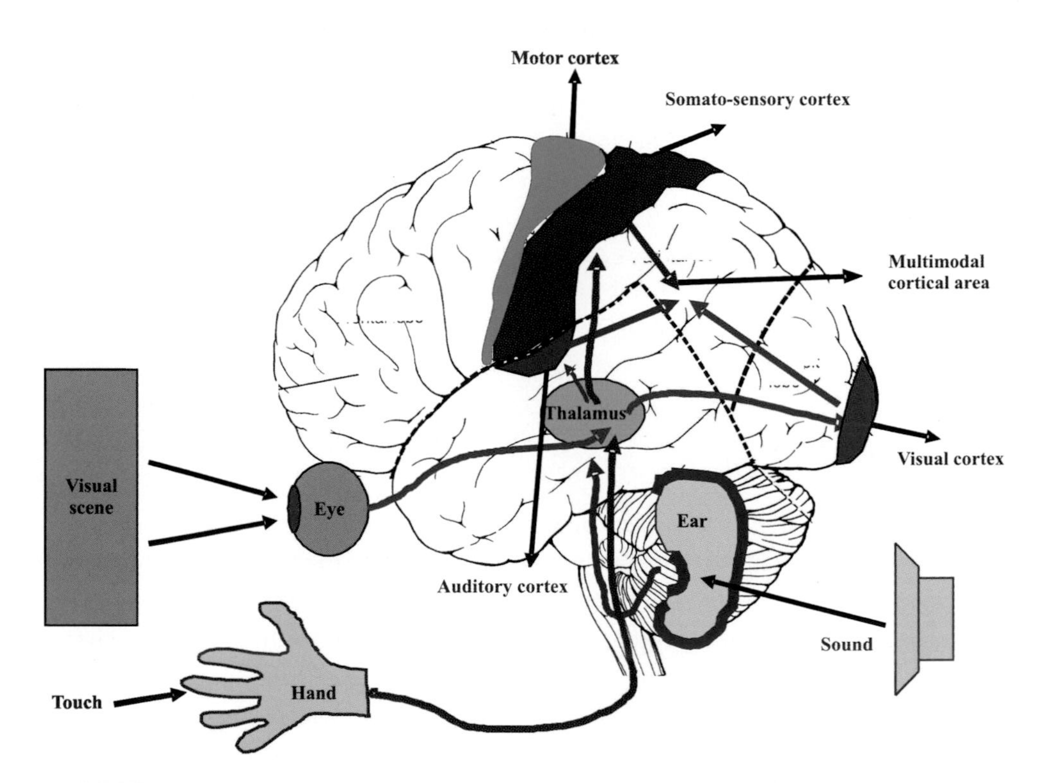

Figure 6.1. For most sensory modalities (visual, auditory, somatosensory), peripheral information comes first in the thalamus nuclei and then to cortical areas specific to each modality. In the cortex, the part of the brain that evolved later in mammals, specialized areas, referred to as unimodal, process information which is specific for each modality and are classified in terms of primary and secondary areas depending on the complexity of processing (primary areas: auditory, A1; visual, V1; somatosensory, S1). There are also multi-modal cortical areas, which integrate information from multiple sensory modalities. Finally, the cortical motor areas organize the motor command to the muscles.

2 Information coding

Different models describe how neuronal assemblies encode sensory and motor information. The first is based on the frequency of the firing of action potentials by neurons. This theory describes neurons as integrators. They integrate their synaptic inputs and discharge action potentials with a frequency depending on the input intensity. The specific characteristics of the stimulus are encoded by the combination of the discharge rate of a group of neurons. An imprecise timing of action potentials in a cell is allowed because only the average frequency of discharge matters, measured on a relatively long timescale (tens of ms). Thus, the detailed temporal course of action potential firing and the temporal correlations between discharges of different neurons convey no additional information. In a famous experiment by Georgopoulos [1] (Fig. 6.2), the discharge of a neuron in the primary motor cortex of a monkey is measured during the performance of a motor task, which consists in moving a joystick according to the position of a target on a screen. In these experiments, neurons have a firing rate that varies with the angle of displacement of the joystick (Fig. 6.2a). Therefore, for each neuron, we can define a tuning curve, which is the average firing rate as a function of the angle of the joystick. This curve possesses a maximum at the "preferred angle" of the neuron (Fig. 6.2b). Measuring the discharge frequency of that neuron provides the knowledge of the angle of the movement. However this measurement across a single neuron predicts the angle with a low accuracy. A much more accurate estimate can be obtained by taking into account the response of the population of neurons (Fig. 6.2c). For one direction of movement, we define a "population vector" $\vec{V}(\vec{D})$ as follows: $\vec{V}(\vec{D}) = \sum_i f(i, \vec{D})\vec{V}_i$, where $\vec{V}_i$ is the unit vector pointing in the preferred direction of the neuron i and $f(i, \vec{D})$ is the firing rate of the neuron i corresponding to a movement in the direction $\vec{D}$. Measuring this vector population predicts more precisely the direction of movement. There is a population coding of the movement direction and the encoding is represented by the firing rate of neurons. Studies in the hippocampus [2] or in the somatosensory cortex [3] have provided further evidence of the generality of this type of frequency coding.

However, the frequency code has limitations, because it requires time to measure the firing frequency, as it implies counting several action potentials. The speed at which successive information can be represented by a frequency coding is therefore limited. To overcome this time limit, another possible dimension for coding has been proposed recently. At a given discharge rate, the precise temporal structure of the trains of action potentials, relative to those of other neurons in the population,

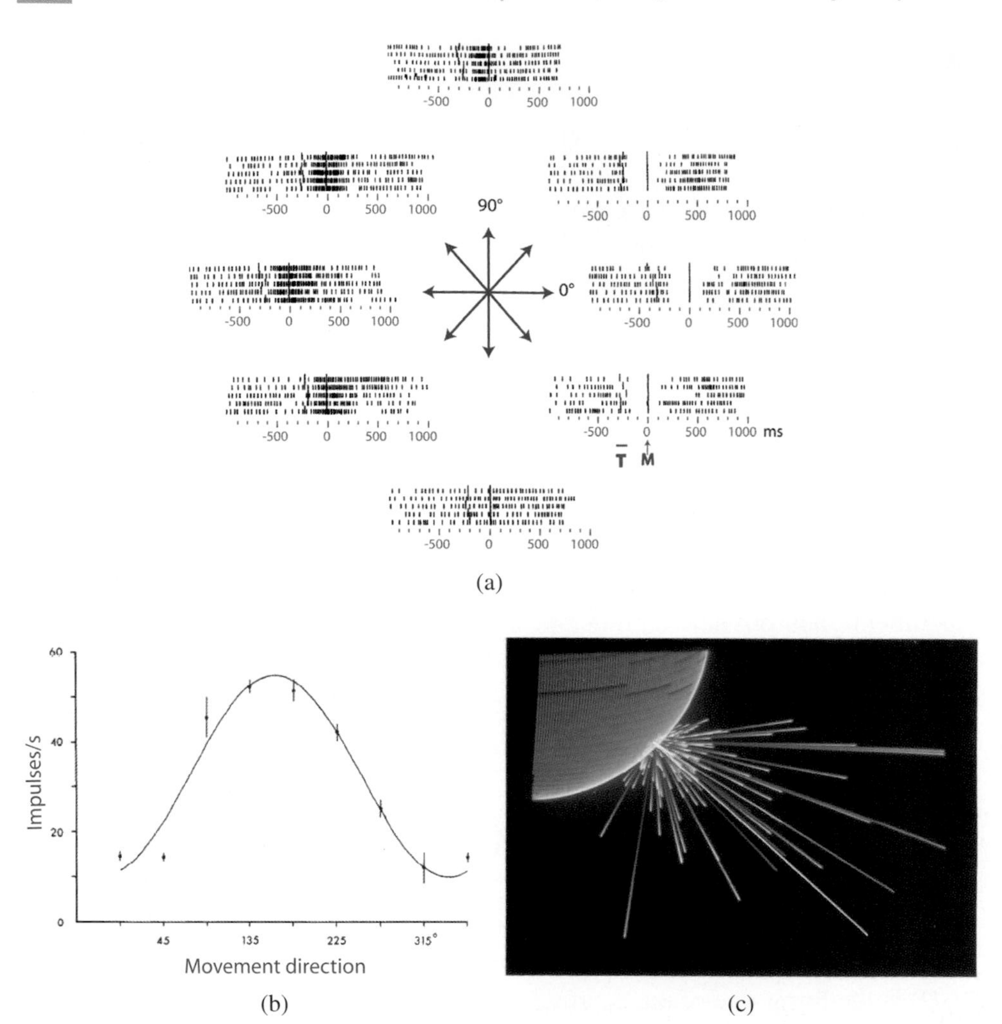

Figure 6.2. Frequency coding. (a) Changes in the frequency of the discharge of a motor cortical cell with the direction of movement. For each direction, five tests are shown (T = 0 corresponds to the beginning of the movement and each vertical bar to an action potential). (b) Tuning curve (discharge frequency) to the direction of movement. A sinusoid is fitted to the curve, which is maximum for the preferred direction of that neuron. Taken from [1]. (c) Population coding of the direction of movement. The blue lines represent the vector contribution $f(i, \vec{D})\vec{V}_i$ of each neuron in the population (N = 475). The direction of movement is indicated in yellow and the direction of the population vector in red. Taken from Georgopoulos *et al. J Neurosci* 1988; 8(8): 2928–37).

may also contain some information [4, 5]. The assumption is that the information is represented by the combination of neurons virtually grouped by the synchronization of their action potentials (Fig. 6.3a). Numerical simulations [6] showed that if enough neurons synchronize their discharges (typically 50 to 100), they may in turn

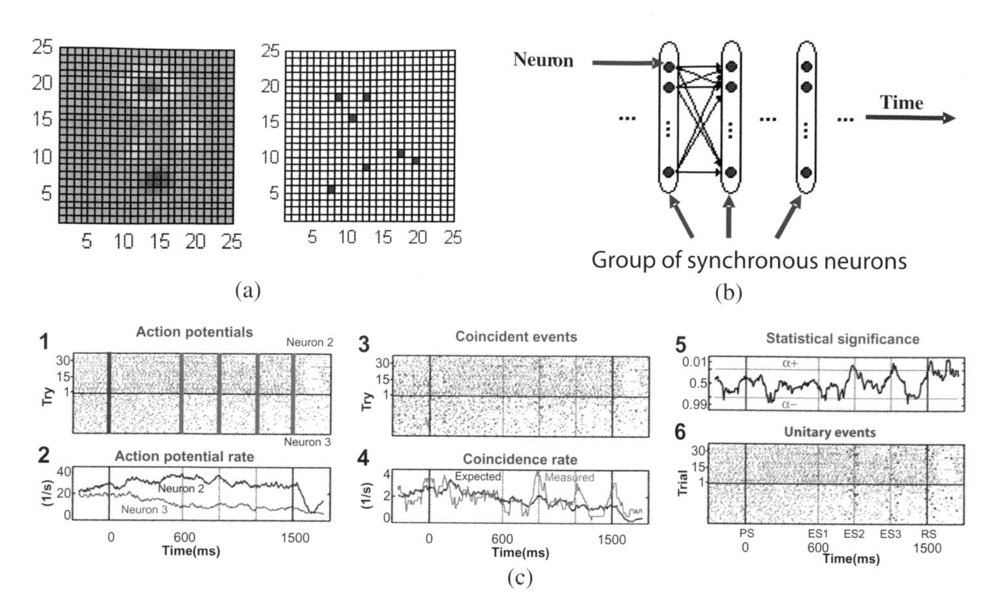

Figure 6.3. Temporal Code. (a) Diagram showing the difference between a frequency code and a temporal code. Each element of the grid represents a neuron belonging to a network with $25 \times 25 = 625$ cells. In the frequency code (left), color indicates the discharge rate of the neuron (blue 1 Hz, 30 Hz red). Each information is represented by the combination of the discharge frequencies of all neurons in the network. In the temporal code (right), the precise temporal relationship between the discharges of neurons in the network also contains information. At a given time, neurons firing an action potential synchronously (marked red) form a subgroup containing specific information about the stimulation. (b) Over time, the subgroup (cigar) of synchronous neurons (black circle) changes and forms a temporal chain of synchronous neurons. This cascade of synchronous neurons represents the information in the temporal code. (c) Change in synchrony during a motor task. 1: The motor task starts at the time indicated by a red line. At the times indicated by a blue line (and only at these times), the monkey may receive an indication asking for movement. The figure shows for 30 trials action potentials (black dots) fired by two recorded neurons. 2: The rate of discharge of the two neurons showed no significant increase at the moments where the monkey may have to perform a movement. 3: The action potentials occurring synchronously between the two neurons are shown in blue. 4: Measurement of the rate of coincident events over time (blue curve). The theoretical rate of expected events is shown in black, if trains of action potentials are Poissonian and uncorrelated between the two neurons (this rate varies when the instantaneous firing rate varies). 5: Detection of coincident events that are statistically significant. 6: Significant coincident events are indicated in red. They occur only at times when a movement is possible. Inspired by [9].

trigger action potentials in a second group of target neurons and so on (Fig. 6.3b). This spread of synchronous activity as a "chain of synchronous neurons" defines a unique path of the neuronal activity and could be a characteristic signature of a given information: different signals activate different synfire chains. The neuron is in this case a coincidence detector: the duration during which it integrates its synaptic inputs is short, of the order of 5–10 ms for a cortical neuron [7]. In recent years, evidence has accumulated showing situations where synchronization of neurons is correlated with behavior [8–10] (Fig. 6.3c).

The temporal and frequency codes show the need to record the activity of large populations of neurons to understand how sensory and motor information are represented and processed in neural networks. The ability to simultaneously record the activity of cells is necessary to access the temporal correlations between neuronal discharges. The vast majority of electrophysiological measurements was obtained with anesthetized or awake animals (mice, rats, cats, monkeys). By inserting a thin metal electrode, it is possible to measure the action potential emitted by an individual neuron, provided that the electrode is positioned adjacent to the cell body, in order to distinguish its contribution from those of other adjacent neurons. The introduction of a quadruplet of electrodes called tetrodes is used to identify the individual contributions of many neurons simultaneously in a sort of triangulation. By inserting a few tens of tetrodes, this approach allowed the recording of the individual activity of dozens of neurons simultaneously *in vivo*, in animals performing behavioral tasks. However, electrophysiological techniques suffer from some limitations. First, they record indiscriminately populations of neurons of heterogeneous cell type. In addition, the relative spatial location of recorded cells is known with a very low accuracy. Finally, the recorded cells are relatively far from each other and the comprehensive recording of neurons constituting a local network is impossible, because it is extremely difficult and invasive to insert many tetrodes in a restricted area. Instead, optical microscopy could provide a precise spatial mapping of brain activity, with the added benefit of identifying the cell types.

3 Optical recordings of neuronal activity

3.1 Scattering and absorption

Biological tissues absorb light. The absorption is characterized by an exponential decay of the light intensity, with a characteristic length l_a, which depends on the wavelength λ and is typically of the order of 1 mm for visible and near infrared light. At short wavelengths (blue or UV), some amino acids, DNA and proteins strongly absorb, while water dominates the near-infrared absorption beyond 1 μm. In the blue-green, hemoglobin has a strong absorption, making the observation of tissue irrigated by the blood particularly difficult. Among the photons that are not absorbed, we can distinguish two kinds. i) Ballistic photons that propagate following the Fermat's principle: along a straight line in a homogeneous medium, along a curved trajectory defined by the variations in the optical index at spatial scales much larger than λ. ii) Scattered photons, which follow random paths. To obtain images, one should ideally be using ballistic photons, because images are blurred by scattering, thus reducing their contrast and resolution. In tissues, composed of objects of sizes covering a wide range of scales, scattering is both due to very small objects as compared to λ, acting as isotropic scatterers (Rayleigh regime) and to large objects of the order of λ or more, radiating mainly in a direction close to the incident direction (Mie regime). Scattering in a heterogeneous environment is characterized by two quantities, the scattering length $l_s(\lambda)$, which is the average distance between two scattering events and the scattering anisotropy g defined $g = \langle \cos(\theta) \rangle$, where θ is the scattering angle and $\langle\rangle$ is the mean over many events. In biological tissues, scattering occurs mainly in the direction of the incident beam ($g \sim 0.8-1$), with two important properties:

(i) The propagation of light is more affected by scattering than by absorption: in the visible spectrum, the characteristic length scale of scattering is much smaller than the characteristic absorption length scale, $l_s \sim 50\ \mu\text{m} \ll l_a$.
(ii) l_s increases with λ: infrared light penetrates deeper than visible light in tissue before being scattered ($l_s \sim 200\ \mu\text{m}$ for $\lambda = 1\ \mu\text{m}$).

Imaging techniques will be classified into two families:

(1) Those using ballistic photons, which have a high spatial resolution to distinguish individual cells and their sub-compartments, but with low tissue penetration of the order of a few l_s. They usually benefit from the transparency window for absorption and scattering in the near infrared to increase the maximum depth of imaging.
(2) Methods that use scattered photons, which provide low spatial resolution, but which may have a larger penetration depth.

3.2 Optical sensors to measure the electrical activity

We can distinguish measurements that do not require the introduction of specific markers and those that use probes, mostly fluorescent. A first approach uses an optical signal related to the blood flow following the activation of a brain area. When a brain area is activated, there is a local increase in oxygen consumption. Oxygen is supplied locally by blood in the form of di-oxyhemoglobin (HbO_2). An active area is enriched in HbO_2, compared to the hemoglobin unbound to oxygen (Hb). The absorption spectra of the two molecules are different in the visible range (Fig. 6.4a). Intrinsic imaging provides an image of the cortical activity at large scale (several mm) by measuring the reflectivity of the cortex at a wavelength where the absorption spectra of HbO_2 and Hb differ significantly.

This approach is a rather indirect measure of the electrical activity. To obtain functional information on neurons, it is necessary to use signals with a cellular origin. The introduction of fluorescent probes is the main approach. Many optical techniques are based on fluorescence, including "traditional", confocal and two-photon fluorescence microscopy. There are at least two families of fluorescent probes to image the electrical activity of neurons. The first consists of probes with fluorescence emission that is modulated by the membrane potential of neurons (VSD, Voltage Sensitive Dyes). These probes are inserted into the membranes of neurons and their fluorescence emission is dependent on the membrane potential (Fig. 6.4b). There are a wide variety of VSD probes. They have usually a fast dynamical response and the fluorescence emission is modulated on the time scale of the action potential (1 ms), with maximum amplitude of variation of the order of only 0.1%. The main advantage of these sensors is that they directly measure the membrane potential, and with an excellent temporal resolution. Their limitations are the small signal changes during an action potential and their photo-toxicity. Currently, the development of protein-based VSD is in progress (Fig. 6.4b). The second group consists of probes with fluorescence that is modulated by the intracellular concentration of calcium. As the membrane of neurons possesses calcium channels with conductance dependent on the membrane potential, each action potential causes a brief influx of calcium into the cell. The resulting increase in intracellular calcium concentration has a rapid rise followed by an exponential decay with a time constant of about 500 ms. This variation in the concentration of intracellular calcium is called action potential-evoked "calcium transient". A fluorescent calcium probe S binds calcium in a reversible chemical equilibrium: $Ca + S \rightleftharpoons CaS$. If the fluorescence of the complex CaS is different from that of the free probe S, the measurement of the fluorescence allows for the measurement of changes in intracellular calcium concentration and therefore the detection of action potentials (Fig. 6.4c). The advantage of these probes is that the changes in fluorescence is of the order of 10% during an action potential. For this reason, most *in vivo* studies at the cellular level use these calcium probes, despite their relatively slow kinetics. After calibration of the shape and amplitude of a typical calcium transient, algorithms can be used to identify the emission times of the action potentials. The temporal precision at which action potentials are identified varies between 10 ms and a few tens of ms. Two types of probes are used: "chemical" probes, which are injected with a pipette locally in the brain, which allow for the transient labeling of extended neural networks, and protein-based probes which are most often injected in viral form. The latter protein-based probes open the possibility of labeling that are chronic and specific of a cellular subtype.

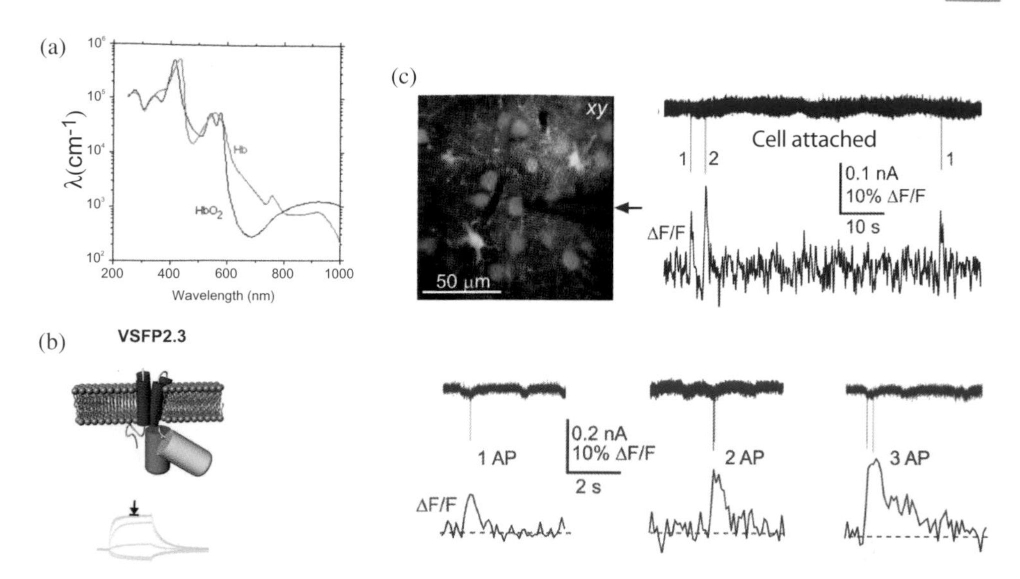

Figure 6.4. (a) The absorption spectra of hemoglobin Hb and di-oxyhemoglobin HbO_2. The difference of the absorption spectra is used in intrinsic imaging to visualize the local activation of the cortex. (b) Example of a protein-based fluorescent voltage probe. It consists of a voltage-sensitive membrane domain and two intracellular domains operating on the principle of FRET (fluorescence transfer mechanism, which is very sensitive to the inter-domain distance). A conformational modification of the probe due to a change of potential changes the distance of the two domains and therefore the fluorescence (see traces below: responses to a voltage step during 500 ms from an intial voltage of −70 mV to potentials of −140 to +60 mV. From Mutoh *et al. Plos One* 2009; 4: e4555 2009). (c) Example of a signal measured with a chemical calcium fluorescent probe. Top left: image (two-photon microscopy) of neurons in the rat cortex labeled with the calcium sensor. Arrow: micropipette used to measure the potential of a neuron. Top right: simultaneous measurements of membrane potential and fluorescence. Bottom: zoom on electrophysiological and calcium responses corresponding from left to right to 1, 2 or 3 action potentials emitted by the cell. Taken from Kerr *et al. PNAS* 2005; 102(39):14063–8).

4 Functional organization of the cortex at the level of a cortical column

A fascinating feature of the cortex is that neurons with the same selectivity for a sensory parameter tend to cluster locally, forming a functional map. These maps were first highlighted in electrophysiology, but with a limited sampling of space. It was therefore difficult to analyze the fine organization of the cortex on spatial scales smaller than a few hundred microns. Intrinsic imaging addresses this issue from a new perspective [13]. This technique uses as optical signal the reflectance changes associated with the activity of a brain area (see Section 3). A series of parameters of sensory stimulation (e.g. the position of an object in the visual field, the angle of a bar, etc.) are presented and functional maps are obtained showing the existence of areas with a selectivity for a particular value of each parameter of the stimulation. Note that we observe only the 2D projection of the organization, because this technique, which uses ballistic and scattered photons, has no axial resolution along the optical axis of the microscope and a lateral resolution of a few tens of microns. Mappings have been observed in many sensory cortical areas. The area that has been most studied is probably the visual cortex. Figure 6.5b shows a

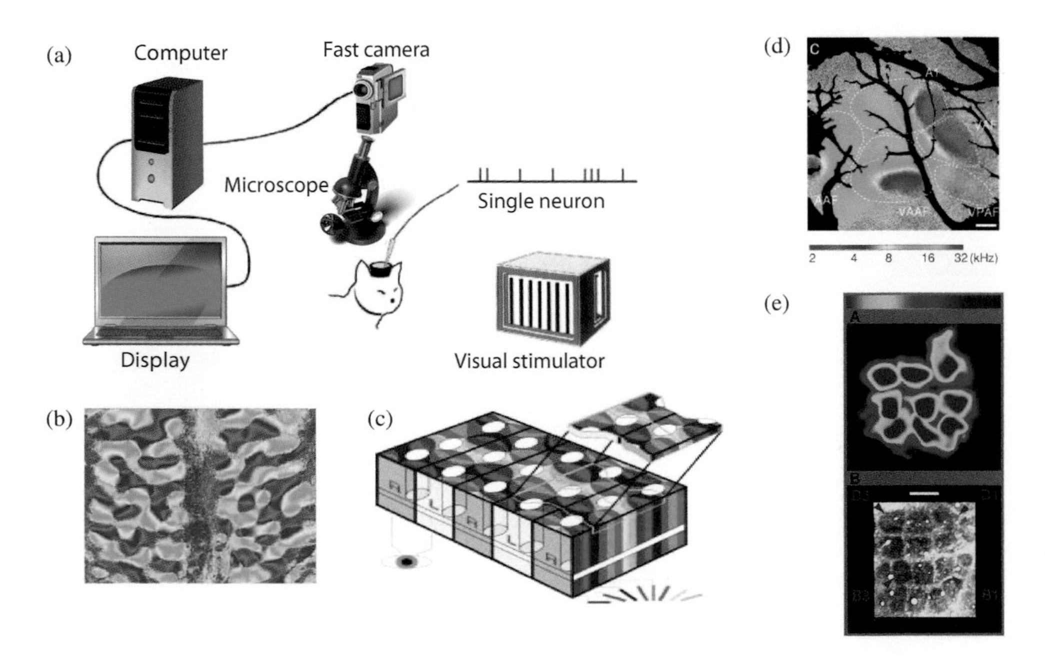

Figure 6.5. Functional organization of cortical areas revealed by intrinsic optical imaging. (a) Principle of intrinsic imaging (taken from Tsodyks *et al. Science* 1999; 286: 1943). (b) Areas selective to the orientation of bars in the visual cortex. The singular points around which areas are organized in spirals are called "pinwheel" (From Kalatzky *et al.* (2003) *Neuron*, 38, 529–545). (c) Schematic of the superposition of maps obtained for different parameters of the visual stimulation. Black lines: limits of ocular dominance columns containing neurons with selectivity for one eye. White ovals: groups of neurons responsible for the perception of color. Pinwheels: organization of areas responding preferentially to a particular orientation. The zoom is a basic module ($400 \times 800\,\mu$m), which contains about 60,000 neurons dealing with these three characteristics of the stimulus (orientation, depth, color). Taken from Grinvald A. *et al.* (1999) in Modern Techniques in Neuroscience research. U Windhorst and H. Johansson (eds), Springer Verlag. (d) Auditory Cortex: tonotopy map (selectivity to the sound frequency) in the primary auditory area A1 and secondary auditory areas (From Kalatzky *et al. PNAS* 2005; 102(37): 13325–30). (e) Somatosensory cortex: map of the response to the stimulation of nine facial macro-whiskers of rats and comparison with the anatomical map of the barrels (from Masino *et al. PNAS* 1993; 90(21): 9998–10002).

map for the orientation of bars, which are involved in the detection of an object's contour. Other maps have been identified, such as those for the position of an object in the visual field or for ocular dominance (responsible for depth discrimination). Maps observed for different parameters of the sensory stimulation overlap spatially. In Figure 6.5c a scheme of the functional organization of the primary visual cortex shows the relative spatial arrangement of maps for orientation, ocular dominance and color. Functional maps were observed in most sensory areas. In the auditory system (Fig. 6.5d), cortical neurons show selectivity to sound frequency. Spatial segregation is observed in iso-frequency bands. It is quite remarkable that this spatial organization actually extends beyond the primary auditory cortex, to the secondary auditory cortical areas involved in more complex representations of the auditory scene. The somatosensory system also has a functional mapping. Rodents have long facial vibrissae (whiskers) involved in the recognition of objects, i.e. the detection of their position and their shape and texture. In the primary somatosensory cortex S1, neurons responding preferentially to a specific whisker deflection are organized in areas called "barrels" (Figure 6.5e).

The universality of the observation of a large number of functional maps in many species and sensory modalities raises the question of their functions. They might facilitate the emergence of high selectivity: the presence of strong recurrent connectivity between close neighbor neurons tends to increase the selectivity of neurons belonging to homogeneous areas. In addition, they might be useful to the organization of calculations or associations involving several parameters through the superposition of several functional maps, which facilitate connections between these parameters.

5 Microarchitecture of a cortical column

To observe the functional organization of the cortex at the cellular level, it is necessary to have an optical technique with a 3D diffraction-limited resolution ($\sim\mu$m) in scattering biological tissues. As intrinsic imaging uses visible light, it does not penetrate deeply into the tissues. Similarly, confocal laser scanning microscopy uses scattered visible light for fluorescence excitation. In addition, the emitted fluorescence photons are themselves scattered and thus filtered by the confocal pinhole. Thus, confocal imaging does not allow imaging of neurons in the cortex at depths larger than that of a few tens of microns.

To limit the effects of scattering a solution was proposed in the early 90s, which brought a revolution into neuroscience: two-photon fluorescence microscopy (TPFM) [14, 15]. Its principle is as follows (Fig. 6.6a). The idea is to use a fluorescence excitation in the near infrared to increase the penetration depth of photons into

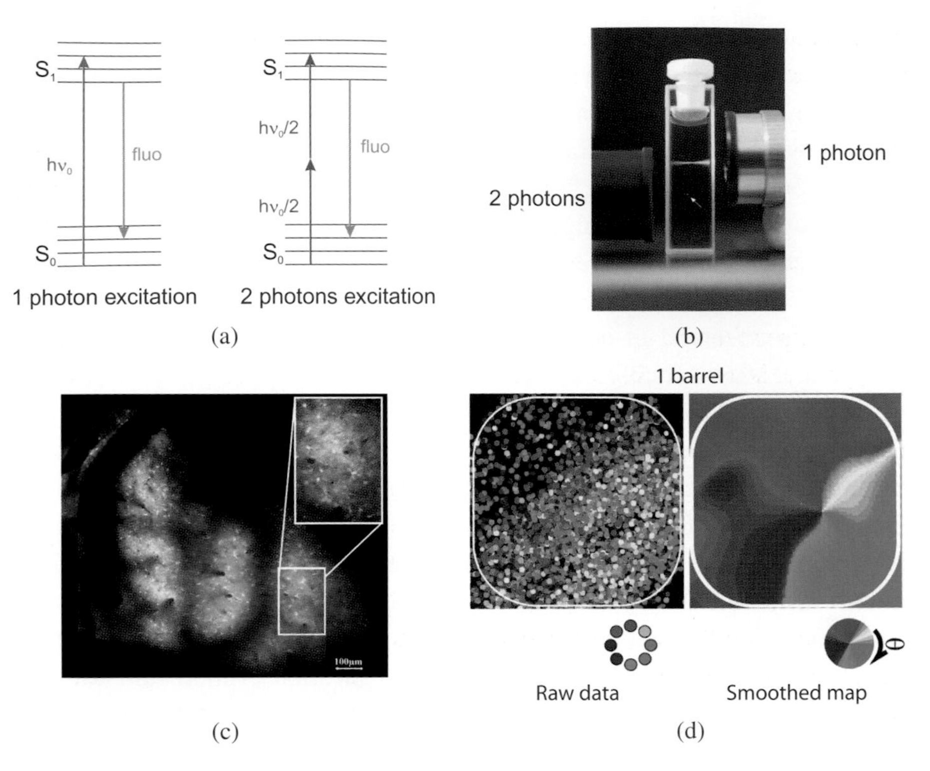

Figure 6.6. (a) Principle of two photon fluorescence microscopy. A nonlinear absorption transition of two infrared photons excites the fluorescence with a wavelength less scattered by the tissues. (b) The fluorescence excitation is located at the focus of the microscope. Top: one-photon excitation, which is not localized; bottom, two-photon excitation, localized (image produced by Brad Amos, LMB Cambridge). (c) *In vivo* two-photon microscopy images of glial cells (orange) and neurons (green). All cells were labeled with a calcium probe emitting in the green and glial cells were specifically labeled with a red probe. (d) Map of selectivity to the direction of deflection of a whisker. White contour: limit of one barrel. Left: individual neurons are spatially separated according to their preference for the direction of deflection of the whisker (the angles are colour-coded, see insert). Right: a spatially-smoothed map shows the average organization ("pinwheel").

the tissue, limiting scattering. But the absorption spectrum of most fluorophores is in the visible. To be able to excite fluorescent probes with an infrared photon of low energy, transition of the ground state to the excited state of the fluorescent molecule is obtained by simultaneous absorption of two photons of low infrared energy. After non-radiative de-excitation, a visible photon is emitted (Fig. 6.6a). This is a nonlinear effect, because it depends on the simultaneous absorption of two photons and the fluorescence intensity varies quadratically with the excitation intensity. For the two-photon transition to occur with a significant probability, laser radiation is temporally concentrated in high-energy pulses, which have a duration of ~100 fs and are focused spatially by a high numerical aperture objective. As the transition is nonlinear, it occurs only in the immediate vicinity of the focus of the lens (Fig. 6.6b). Therefore, the fluorescence is emitted in a volume of space of diffraction-limited size and it is not necessary, as in a confocal microscope, to filter the fluorescence light emitted at the focus of the lens by a confocal pinhole. All the fluorescence is imaged onto the photo-detector. An image is then reconstructed, such as in confocal microscope, by scanning of the laser beam spot in the sample.

Knowledge of the photo-detector signal and the position of the laser in the optical field over time is used to "reconstruct" the image in a computer. As the fluorescence excitation is obtained by ballistic photons, the spatial resolution is high, diffraction-limited. As all fluorescence photons, ballistic or scattered, are detected, this image is bright, even at large depth. Two-photon microscopy allows us to image deep into biological tissues, particularly in the brain, to a depth of 500 μm to 1 mm. Images of neurons can be obtained *in vivo* in the brain of an anesthetized animal (Fig. 6.6c). A craniotomy is still necessary, but the technique is minimally invasive to the extent that only the photons penetrate the tissues! Using this technique, new updated versions of cortical functional maps could for example be developed. A first study looked at the map of orientation selectivity in the visual cortex of the cat. The observation of the average map obtained by intrinsic imaging (Fig. 6.5b, top) was analyzed at the single cell level. This study showed that the pinwheel organization was observed at the cellular scale. This observation raises fascinating questions: knowing that these cells have extensive dendritic trees of several hundred microns, they integrate information from neurons with very different selectivities. The origin of the difference in selectivity of the neighbour cells close to a pinwheel remains a mystery!

Two-photon microscopy was used to analyze the organization of functional maps in other sensory modalities, such as auditory and somatosensory. In the auditory cortex, tonotopic observation areas at the cellular level showed that the observed intrinsic imaging bands masked a strong disorder at the cellular level. In the case of the somatosensory cortex, two-photon microscopy was used to analyze the functional organization within a single barrel. The presence of areas selective to the direction of whisker deflection and radially arranged in a barrel was demonstrated. These subdomains emerge with the age of the animal and with increased sensory experience (Fig. 6.6d and [12]).

6 Dynamics of neuronal populations

We have shown in the introduction that one of the ultimate goals of integrative neuroscience is to correlate the unit activity, i.e. activity of individual cells, with the properties of sensory stimulation or motor control, to understand the neural code. Intrinsic imaging and two-photon microscopy have allowed us in recent years to better understand the basic principles of the functional organization of the cortex. Optical imaging also proves to be a valuable tool for studying the dynamics of neuronal populations and coding of information by neural networks. Two-photon microscopy was used to measure the response of neurons averaged over a large number of repeated stimulations to measure the selectivity curve of each neuron. From the selectivity curve, the preference of each neuron was determined and the associated functional map was obtained thanks to the unique possibility of spatial localization of neurons offered by this technique. It is also possible to study the response of neurons evoked by each stimulation, action potential by action potential. This approach has been carried out not only on anesthetized animals, but also on awake animals. In Fig. 6.7, we show a configuration where the animal is kept head-fixed under the microscope, the body being carried by a free spinning ball. Moreover, since the animal can move substantially freely, the rotational movement of the ball can be used to recover the movement of the animal. If it is placed in an appropriate virtual environment, it is possible to follow the cortical activity when it acts as if it moves freely (Fig. 6.7a). In the experiment of Fig. 6.7, the mouse is trained to run in response to a signal (a puff of air on its skin). It is possible to track the individual activity of large neuronal populations during periods of rest and motion. The apparent speed of movement of the mouse is measured from the rotation of the ball. One can observe that the dynamics of neurons is strongly correlated with that of the race. For example neurons 1, 2, 3, 4 have discharges systematically correlated to the duration of the race, while other neuron responses are more poorly correlated (Fig. 6.7c). On the other hand, we observe that the correlations between neural activities are high for neurons with activity that are strongly correlated with the race, but are lower for the other (Fig. 6.7d). This experiment opens fascinating perspectives for the understanding of information coding. It will probably be possible to test which of the two encoding schemes, time or frequency, better predicts the behavior of the animal. It will also be possible to analyze the fraction of neurons involved in the representation of information: is it focused on only a few cells, on the entire network or a part of it? These questions fascinate neurobiologists for a century and their answers are readily available.

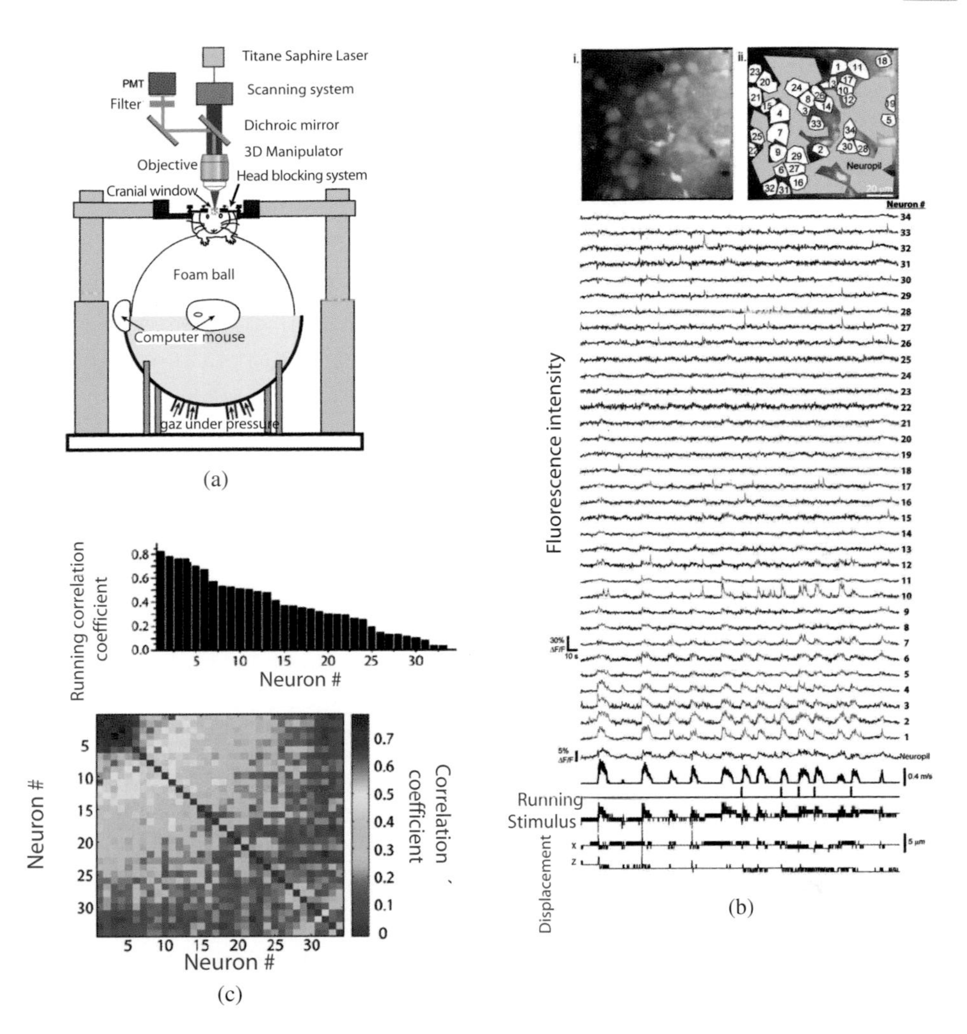

Figure 6.7. (a) *In vivo* two-photon microscopy in the somatosensory cortex of the hind paw of awake mice, maintained head-fixed. (b) Top: image with a calcium dye; detected neurons are shown on the right image by numbers; the neuropil (symbolized by gray area between the cell bodies of neurons) is also indicated. Bottom: fluorescence traces for the 34 neurons shown in the upper figure and for the neuropil. The running speed, the stimulation of the leg by an air puff and the residual movement of the brain are shown. Fluorescence transients are indicated in red. (c) Top: correlation coefficient for each neuron between neuronal activity and the running of the mouse, bottom: correlation of activity between neurons. Taken from Dombeck *et al. Neuron* 2007; 56: 43–57).

7 Outlook

The functional organization of the cortex and the representation of information by neural networks have been described much more accurately thanks to the recent progress in optical methods. These experiments are complex and still suffer from many limitations in terms of signal extraction from the experimental noise, temporal resolution, penetration depth or analysis of cell types. Many teams around the world are working to exceed these limits.

Faster. It is crucial to improve the temporal precision of optical measurements of electrical activity, especially for the analysis of the role of synchronies in encoding of information. On one hand, indirect measurement of action potentials through calcium probes limit the precision to a few tens of ms, given the kinetics of calcium transients. The increase in the amplitude of the signal for VSD probes would certainly be a major asset. On the other hand, improving temporal resolution also involves significant efforts to setup new strategies for scanning the laser. The ultimate noise is the photon noise whose influence can be reduced by increasing the time spent on each cell body of neurons. This can be achieved by considering point-by-point scans, one cell body to the next. For this purpose, the replacement of traditional oscillating mirror galvanometers by acousto-optic deflectors proves to be an extremely promising solution. These deflectors diffract the light wave in a given direction, which can be changed in a few microseconds. By this method, the timing of an action potential was measured with an accuracy of about 10 ms, bridging the gap between electrophysiology and optics.

Deeper. Currently, imaging techniques allow access to neuronal populations located up to 500 microns deep, limiting their use to the outer part of the external structures of the brain (cortex, cerebellum). To go deeper, two strategies have been followed. First, biological tissues have a heterogeneous structure creating spatial fluctuations of the refractive index at large spatial scales. These fluctuations are responsible for optical aberrations, such as those caused by the atmosphere in astronomy when observing a star. Imitating the methods developed in astronomy over the past 20 years and which have revolutionized the telescopes, researchers are now trying to combine adaptive optics and two-photon microscopy to restore a resolution only limited by diffraction in deep tissue and also much brighter images, increasing the maximum depth at which imaging is obtained. A second approach to increase the penetration depth is to develop endoscopic two-photon microscopes through the use of optical fibers and gradient index lenses. Ultimately, it is hoped that these studies will not be limited to the outer layers of the cortex, but can be applied also to other internal structures such as the hippocampus or thalamus.

More specific. The use of genetic markers can also be used to identify the cell types involved in a given activity pattern. Coupled to the opto-genetic methods recently developed, which allow the control by optical methods of the specific activity of certain neuronal subpopulation, these advances suggest that we can eventually dissect the specific role of each cell subtype in large neuronal populations.

Bibliography

[1] Georgopoulos AP, Kalaska JF, Caminiti R, Massey JT. On the relations between the direction of two-dimensional arm movements and cell discharge in primate motor cortex. *J. Neurosci* (1982) **2**(11), 1527–37.

[2] Wilson MA, McNaughton BL. Dynamics of the hippocampal ensemble code for space. *Science* (1993) **261**(5124), 1055–8.

[3] Nicolelis MA, Baccala LA, Lin RC, Chapin JK. Sensorimotor encoding by synchronous neural ensemble activity at multiple levels of the somatosensory system. *Science* (1995) **268**(5215), 1353–8.

[4] Abeles M. *Corticonics neural circuits of the cerebral cortex*, CUP Cambridge (1991).

[5] Abeles M, Bergman H, Margalit E, Vaadia E. Spatiotemporal firing patterns in the frontal cortex of behaving monkeys. *J. Neurophysiol* (1993) **70**(4), 1629–38.

[6] Diesmann M, Gewaltig MO, Aertsen A. Stable propagation of synchronous spiking in cortical neural networks. *Nature* (1999) **402**(6761), 529–33.

[7] Leger JF, Stern EA, Aertsen A, Heck D. Synaptic integration in rat frontal cortex shaped by network activity. *J. Neurophysiol* (2005) **93**(1), 281–93.

[8] Vaadia E, *et al.* Dynamics of neuronal interactions in monkey cortex in relation to behavioural events. *Nature* (1995) **373**(6514), 515–8.

[9] Riehle A, *et al.* Spike synchronization and rate modulation differentially involved in motor cortical function. *Science* (1997) **278**(5345), 1950–3.

[10] Braitenberg V, Heck D, Sultan F. The detection and generation of sequences as a key to cerebellar function: experiments and theory. *Behav. Brain Sci.* (1997) **20**(2), 229–45; discussion 245–77.

[11] Ohki K, *et al.* Highly ordered arrangement of single neurons in orientation pinwheels. *Nature* (2006) **442**(7105), 925–8.

[12] Kremer Y, Léger JF, Goodman D, Brette R, Bourdieu L. Late emergence of the vibrissa direction selectivity map in the rat barrel cortex. *J. Neurosci* (2011) **31**(29), 10689–700.

[13] Ts'o DY, Frostig RD, Lieke EE, Grinvald A. Functional organization of primate visual cortex revealed by high resolution optical imaging. *Science* (1990) **249**(4967), 417–20.

[14] Denk W, Strickler JH, Webb WW. Two-photon laser scanning fluorescence microscopy. *Science* (1990) **248**(4951), 73–6.

[15] Kerr JN, Denk W. Imaging *in vivo*: watching the brain in action. *Nat. Rev. Neurosci.* (2008) **9**(3), 195–205.

7

Physical principles of hearing

Pascal Martin, *CNRS researcher, Laboratoire Physico-Chimie Curie Institut Curie and CNRS, Paris.*

1 Psychophysical properties of hearing

To a first approximation, the ear works as a biological analogue of a microphone: it responds to sound-evoked mechanical vibrations by producing electrical signals that travel along the auditory nerve to the brain (Fig. 7.1). Psychophysical experiments from the beginning of the twentieth century have revealed that the ear is a sound detector endowed with remarkable physical specifications.

1.1 Sensitivity, frequency selectivity and dynamic range of auditory detection

The ear displays exquisite sensitivity. The faintest sounds that we can detect correspond to a deviation of the air pressure from the mean atmospheric pressure (1 atm $= 10^5$ Pa) of only 20 μPa and elicit vibrations of the tympanic membrane of less than one picometer. This amplitude is comparable to that of the Brownian movement that shakes our ears even in the absence of any sound stimulus! The ear thus operates at a physical limit imposed by its own thermal fluctuations. Interestingly, the ear is less sensitive to sounds that last less than about 200ms; longer sounds are easier to detect. In 1948, the physicist Thomas Gold recognized in this property the signature of a mechanical resonance. By accumulating mechanical energy from cycle to cycle, the amplitude of sound-evoked vibrations would grow until the threshold of detection is reached. By modeling the ear as a harmonic oscillator

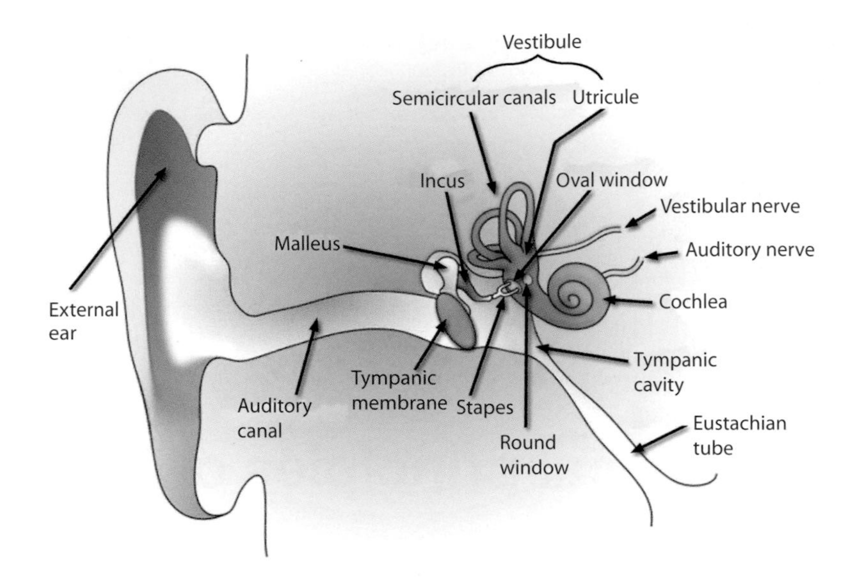

Figure 7.1. The human ear. Sound pressure waves first set the middle ear into mechanical vibrations. The middle ear is composed of the tympanic membrane (also called the eardrum) across the ear canal, and of a chain of three ossicles (incus, malleus, stapes). This arrangement affords efficient transmission of the acoustic energy that propagates through the air to sensory structures that are immersed in physiological fluid in the auditory organ — the cochlea. The ear also mediates the sense of balance by using the vestibular system to detect head accelerations. In particular, the semi-circular canals respond to head rotations. Adapted from (http://fr.wikipedia.org/wiki/Oreille).

(a mass attached to a spring and subjected to viscous drag), Gold observed that the ear operates as a high-quality resonator, as if vibrations inside the cochlea were very lightly damped by the surrounding fluid. However, the vibrating structures are immersed in a fluid too viscous to allow for passive resonance. To alleviate this apparent contradiction, Gold was the first to argue that the ear must be endowed with an active process of amplification that reduces friction on the inner ear. This conclusion was largely ignored at the time and it took thirty years before the existence of the "cochlear amplifier" was finally accepted.

The ear is also remarkable because it can respond to sounds with frequencies that vary over three orders of magnitude, typically from 20 Hz to 20 kHz in human hearing. Other mammals, like whales and bats can even hear sounds up to 100 kHz. These frequencies imply that mechano-electrical transduction must be fast enough to follow temporal variations of the acoustic pressure over timescales as short as a few tens of microseconds. In addition, the ear can discriminate sounds with sharp frequency selectivity. A trained musician can indeed hear the difference between two sounds, when they are played one after the other, that differ by only 0.2%, corresponding to a couple of hertz for frequencies near one kilohertz.

Finally, from the threshold of hearing to sounds so intense that they hurt, the acoustic pressure spans six orders of magnitude, from 20 μPa to 20 Pa. The acoustic power, which goes like the square of the acoustic pressure, thus varies over twelve orders of magnitude. To cope with such a wide range of sound intensities, we use a logarithmic scale, the decibel. In this unit, multiplying the acoustic pressure P by ten results in an increase by 20 decibels of the intensity I_{dB}: $I_{dB} = 20 \log (P/P_{ref})$, in which $P_{ref} = 20\ \mu$Pa. By definition, the mean detection threshold in human hearing is 0 dB, whereas pain starts at 120 dB. For noisy sounds, the smallest intensity change that we can perceive is about 1 dB. This observation suggests that the subjective perception of sound intensity is compressive, meaning that the perceived loudness of sound increases much slower than the acoustic pressure that (objectively) sets the inner ear into mechanical vibrations.

1.2 Auditory illusions: distortions and masking

The ear, unlike standard microphones, does not work as a high-fidelity sound receiver. As first noted in the 18th century by the Italian violinist Tartini, when subjected to the superposition of two sounds at frequencies f_1 and f_2, one can hear not only these two frequencies — the primaries — but also additional "phantom" tones that are not present in the acoustic input. With a frequency difference $f_2/f_1 \simeq 1.1$, the phantom tone that dominates occurs at $2f_1 - f_2$. For a passive mechanical system, one expects a linear regime of responsiveness at low levels of stimulation, for which the response would faithfully represent the stimulus. Distortions ought to emerge when the stimulus becomes so intense that the response can no longer grow proportionally. A simple Taylor expansion $X \simeq \xi F + \alpha F^2 + \beta F^3$ of the transfer function between the stimulus $F = \bar{F}(\sin(2\pi f_1 t) + \sin(2\pi f_2 t))$ and the evoked movement X indicates that the cubic distortion product at $2f_1 - f_2$ ought to grow with $\bar{F}^3$, at least if the magnitude of the stimulus $\bar{F}$ is not too large. The auditory nonlinearity, however, displays surprising properties. First, the phantom tone is heard even at very low levels of stimulation, as if the ear did not present any linear regime of responsiveness. Second, the distortion increases at about the same rate as the stimulus, so that its relative level remains practically constant at 15–20% over a broad range of sound levels. Being only 15 dB less intense than the primaries, these distortion products have even been used in musical compositions!

The auditory nonlinearity also manifests itself by the phenomenon called "masking", by which the perceived intensity of a test tone diminishes in the presence of a second (masker) tone at a nearby frequency. Masking could be a nuisance, but is in fact useful. Being more effective for weak stimuli, masking can enhance the contrast between (strong) signals and (weak) noise.

2 The cochlear amplifier

The dazzling performances of hearing result in part from the mechanical properties of the cochlea, which is the auditory organ of the inner ear. The cochlea takes its name from the greek word 'kochlos', which means snail, because it clearly resembles the shell of a snail (Fig. 7.1). It is inside the confines of this coiled tube that the process of mechano-electrical transduction happens. Cochlear mechanics has been characterized over the past 60 years by measuring *in vivo* the mechanical vibrations evoked by sound stimuli.

2.1 Cochlear mechanics

The cochlea is split in two compartments by a flexible partition. Although it is represented by a simple line in Fig. 7.2, the partition is in fact a complex structure. In particular, it contains the mechano-sensory cells that report transverse vibrations to the brain (see Fig. 7.9). Remarkably, a sound of a given frequency elicits significant vibrations within a restricted region of the partition only. For faint sounds, this region spans about 200 μm of the 35-mm long cochlear tube. In addition, the position of maximal vibration depends on frequency. Low-frequency sounds elicit vibrations near the cochlear apex, whereas high-frequency sounds evoke vibrations near the base (Fig. 7.2). The cochlea in turn works as an "acoustic prism". It distributes spatially, along the cochlear axis, the energy associated with the different frequency

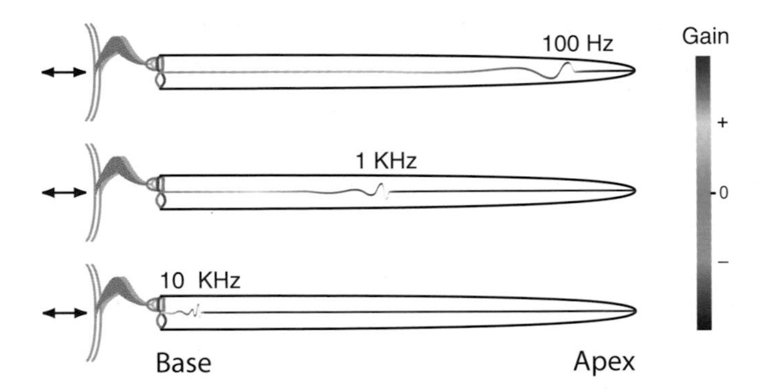

Figure 7.2. The acoustic prism. In this schematic drawing, the cochlea has been unrolled and is represented in a longitudinal cross-section. The cochlear tube is split in two compartments by a "partition". A sound of a given frequency elicits vibrations only within a restricted region of the partition, at a location that varies with frequency. An active process of amplification augments vibration amplitudes with a gain that is here color coded. At any given point, the amplificatory gain depends on the frequency of the acoustic stimulus. The vibration amplitude has been exaggerated by a factor 100,000 with respect to the cochlear length (35 mm in humans). Adapted from Hudspeth A.J., Jülicher F., Martin P., *J. Neurophysiol.* 2010; 104:1219–29.

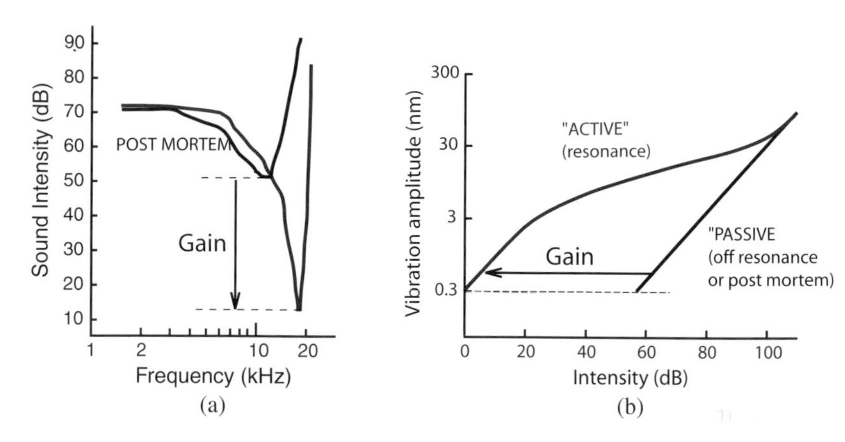

Figure 7.3. The cochlear amplifier. (a) Sound intensity that elicits vibrations of the cochlear partition by 0.35 nm as a function of frequency. In an active cochlea, the relation displays a sharp peak centered at a characteristic frequency, here 18 kHz, of the position where measurements are performed along the cochlear axis. Postmortem, the cochlea is much less sensitive to the sound stimulus, by 40–50 dB, and the resonance is shallower. (b) Plot, in a doubly logarithmic scale, of the vibration amplitude as a function of sound intensity for stimulation at resonance (red line), i.e. at a frequency near the characteristic frequency of the partition at this position, and off-resonance or postmortem (blue line). The dashed line shows the threshold of auditory sensitivity (0.3 nm). The cochlear amplifier lowers the detection threshold by 40–50 dB (gain).

components of a complex sound. Conversely, each cross-section of the partition displays a resonant behavior at a characteristic frequency of the position where the vibrations are measured. The resonance is demonstrated by measuring the sound intensity required to evoke a threshold vibration, typically on the order of one nanometer, as a function of frequency; at the characteristic frequency, the stimulus intensity displays a minimum (Fig. 7.3a). This observation is consistent with the conclusions that Thomas Gold had formulated nearly thirty years earlier on the basis of psychophysical measurements: the cochlea appears to work as a resonator. It is worth noting, however, that the quality of the resonance estimated by Gold is far higher than that evinced by cochlear mechanics. It is likely that auditory processing by the brain plays a key role in the process that sets the frequency selectivity of auditory detection.

For stimulation at resonance, the partition accommodates the six orders of magnitude of sound-pressure levels (0 — 120 dB) that characterize audible sounds within only two to three orders of magnitude of vibration amplitudes (Fig. 7.3b). This compressive nonlinearity results in mechanical distortions of complex sound stimuli that probably explain why two-tone stimuli elicit "phantom tones" in the auditory percept. Notably, even if the partition vibrates by only a fraction of one nanometer ($\simeq$0.3nm) near the auditory threshold, i.e. at 0 dB, linear mechanics would entail vibrations in response to intense sounds (near 120 dB) that would be in the millimeter range. This amplitude is comparable to the diameter of the cochlear tube and is a hundredfold larger than the typical size of a cell. It thus seems natural that some form of compression be associated with the enormous dynamical range of hearing. Auditory distortions in turn appear as a necessary price to pay for being sensitive over a wide range of sound intensities. The response of the partition, noted X, is well described by a power law $X \propto P^{1/3}$ over four orders of magnitude of sound-pressure levels P. Sensitivity, which is defined as the ratio of the response and the pressure, in turn decreases with the power $-2/3$ of the pressure. Because this nonlinearity is present even at very low sound levels, one talks about an "essential" nonlinearity to signify that the linear regime of cochlear mechanics is very restricted, almost non-existent. The compressive nonlinearity that is observed at resonance is in striking contrast with the linear response that is measured at sound frequencies that are much smaller than the characteristic frequency of the resonance. It is intriguing that the cochlear partition behaves as a nonlinear material only when it is simulated near the characteristic frequency.

Notably, the sharp resonant behavior and the compressive nonlinearity that are associated with acute hearing are extremely vulnerable, disappearing after acoustic trauma, anoxia, or death. In these cases, the mechanical resonance is nearly lost (Fig. 7.3a) and the response increases linearly with sound intensity at all frequencies of the sound stimulus (Fig. 7.3b). For this reason, the nonlinear mechanics that is observed in healthy ears is thought to result from a frequency dependent, active amplificatory process. Analysis of the local impedance of the partition supports this inference. Near resonance, the partition indeed appears to be subjected to negative friction. An active motor process must then produce mechanical work to cancel friction forces that impede cochlear vibrations and allow the partition to resonate with the sound stimulus. By amplifying weak stimuli, this active process lowers by nearly 50 dB the minimal sound intensity that elicits vibrations large enough to be detected (about 0.3 nm), which greatly extends the dynamic range of auditory detection.

2.2 Songs in my ears

The ear does not only work as a sound receiver but can also emit sounds, even in the absence of a sound stimulus. These oto-acoustic emissions are usually too weak to be consciously perceived, but can be objectively recorded with a sensitive microphone placed in the ear canal. They provide the most direct evidence that the ear is mechanically active and prove that the ear can sustain mechanical oscillations at auditory frequencies. In some species, the frequency of the emissions can be very high, up to about 63 kHz in the bat. This is remarkable, considering that the corresponding vibrations are produced by the partition, which is a viscoelastic tissue immersed in a fluid as viscous, or more, as water.

Oto-acoustic emissions have been recorded in various vertebrate species, including frogs, lizards, birds, and humans. This suggests that the active process of auditory amplification is ubiquitous and has appeared very early in evolution. Note that some "lower" vertebrates, like the barn-owl for example, are in fact endowed with higher sensitivity and sharper frequency selectivity than many mammals. Mammals, however, differ from other vertebrates by their ability to hear sounds at frequencies above ten kilohertz.

3 Mechanosensory hair cells

The detailed mechanism that would explain the rich phenomenology of auditory amplification is currently the subject of an intense debate. It is however generally agreed that amplification results from active force production by the mechanosensory cells of the inner ear: the hair cells. Hair cells are epithelial cells that take their name from the hair bundle that each cell projects in the surrounding fluid (Fig. 7.4a). The hair bundle is composed of 30–300 cylindrical protrusions — the stereocilia — that are organized in rows of increasing heights and form a staircase pattern that resembles that of organ pipes in churches.

3.1 The hair bundle: a mechano-receptive antenna

Hearing starts with the deflection of the hair bundle. An excitatory deflection, which is directed from the shortest towards the longest stereocilia, evokes a cation influx near the top of the stereocilia, mostly of K^+, but also of Ca^{2+}. This ionic current depolarizes the hair cell, resulting in neurotransmitter release (glutamate molecules) at synapses located near the base of the cell soma. The neurotransmitter in turn prompts an electrical signal that propagates along afferent nerve fibers to which each hair cell is connected. Mechano-electro transduction results from direct mechanical activation of mechano-sensitive ion channels, without intervention of enzymes or secondary chemical messengers. In this scheme, the mechanical stimulus changes the extension of an elastic link — the gating spring– that is connected to the transduction channel.

The stereocilia are interconnected by numerous lateral links (Figs. 7.4b and 7.4c). In particular, an oblique link, called the tip link, runs from the top of each stereocilium to the flank of the nearest taller neighbor (Fig. 7.4c). In response to an external force, each stereocilium pivots about its insertion point in the apical surface of the

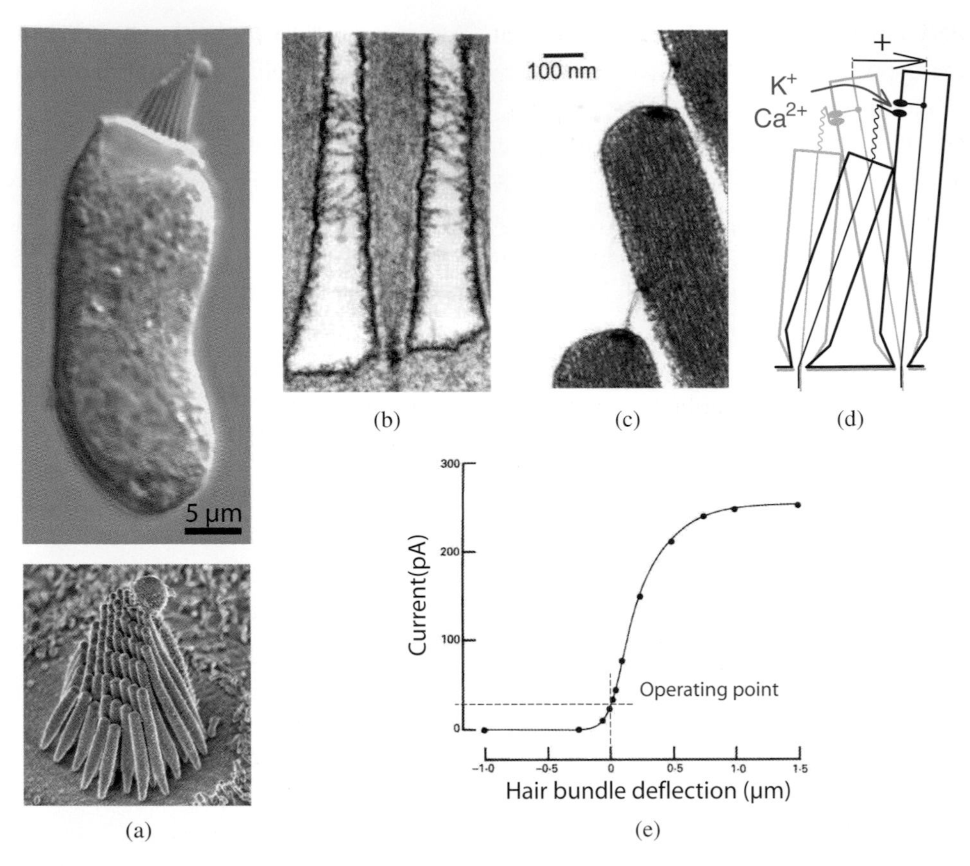

Figure 7.4. Hair cells. (a) Top: an isolated hair cell from the inner ear of a bullfrog (*Rana catesbeiana*). Bottom: the mechano-sensory hair bundle. (b) Pivots. Each stereocilium is an extended, actin-filled villosity that tapers near its insertion into the cell soma. Only a few tens of the ~300 actin filaments that compose the cytoskeleton of a stereocilium enter the cell to form a rootlet. (c) Tip links between the top of one stereocilium and the flank of the nearest taller neighbor. (d) A deflection of the hair bundle in the direction indicated on the figure ("positive" or excitatory direction) increases tip-link tension, which results in the opening of mechano-sensitive ion channels and an increased cation influx of K^+ — the major component of the current — and of Ca^{2+}. (e) Sigmoidal relation between the hair-bundle deflection and the transduction current. The operating point of mechano-electrical transduction is positioned in the steep region of the curve. Pictures (a) (top), (b) and (c) from the Laboratory of Jim Hudspeth (the Rockefeller University, New York, USA) and (a) (bottom) from that of Peter Gillespie (OHSU, Portland, USA). Data in (e) adapted from Crawford C., Evans M.G. and Fettiplace R., *J. Neurophysiol.* 1989; 419:405–434.

hair cell (Fig. 7.4b), resulting in a shearing movement between adjacent stereocilia that modulates tip-link tension. When tension increases, so does the open probability of the transduction channels (Fig. 7.4d). Although the tip link is ideally placed to constitute the gating spring, electron micrographs of the tip link, together with molecular dynamics simulations, reveal a structure that is probably too stiff and inextensible. If the tip link is not the gating spring, a compliant linkage would have to be connected in series to the tip link. In addition to the tip link, horizontal top connectors ensure the mechanical cohesiveness of the hair bundle and the concerted activation of the transduction channels, which is a necessary condition for sensitive detection.

The relation between the hair-bundle deflection and the transduction current follows a sigmoidal curve (Fig. 7.4e). Mechano-electrical transduction is so sensitive that a deflection of only a few tens of nanometers suffices to saturate the transduction current. This deflection is much smaller than the diameter of a stereocilium and corresponds to an angular rotation of the hair bundle of about one degree only.

3.2 Mechanical correlates of mechano-electrical transduction: gating force and negative stiffness

The relation between tip-link tension and the open probability of a transduction channel is reciprocal: if increasing tip-link tension evokes channel opening, then channel opening must reduce tip-link tension. In the following, we will refer to the reduction in tip-link tension upon channel opening as the "gating force" Z (Fig. 7.5a). In a simple two-state model (open/closed) of the transduction channel, one can show that the hair-bundle deflection that results in a change of open probability from 12% to 88% is given by $4\,k_BT/Z$, where k_BT is the thermal energy (Fig. 7.5b). Thus, the larger the gating force, the smaller the deflection to open most of the channels and the more sensitive the transduction process. In practice, measurements of current-displacement relations yield values of $0.25-0.7$ pN for Z; when estimated along the oblique axis of the tip link, the gating force reaches a magnitude of a few piconewtons (2–5 pN).

Gating forces diminish the elastic restoring force exerted by the tip links by producing an extra give in the direction of the applied force. The stiffness of the transduction apparatus is in turn lower than that of the gating spring alone, a phenomenon termed "gating compliance". This effective softening constitutes a fundamental property of any mechano-sensitive system and reflects here an inherent nonlinearity associated with mechano-electrical transduction. The force-displacement relation indeed displays a region of reduced slope over the narrow range of displacements that elicit significant rearrangements of the transduction channels between their open and closed states. It is rather intriguing that gating of only $\sim$50 ion channels can affect the stiffness of a micro-mechanical structure as large as the hair bundle. This is possible because the gating springs contribute a large fraction ($\simeq$80%) of the hair-bundle stiffness. Gating forces can even be large enough (Fig. 7.5b) for

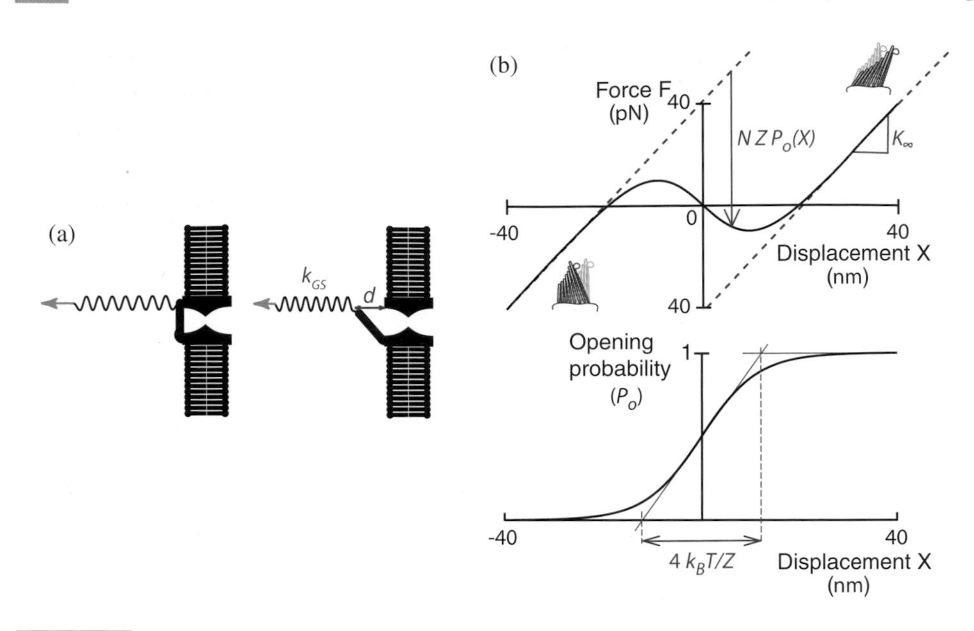

Figure 7.5. Gating compliance. (a) Gating force. The opening of the transduction channel results in a reduction $Z = \gamma\, k_{GS} d$ of tip-link tension (blue arrows), where k_{GS} represents the gating-spring stiffness, d the size of the conformational change associated with channel opening and γ a geometrical projection factor that takes into accounts the fact that gating forces are produced along the oblique axis of the tip link but measured horizontally. (b) Force-displacement relation of a hair bundle. At large negative or positive deflections, at which the transduction channels are either all closed or all open, respectively, and force thus does not change the channels' open probability, the relation is linear. The hair bundle is then equivalent to a simple spring of stiffness $K_\infty \simeq 1$ pN/nm. In contrast, smaller deflections elicit significant rearrangements of the transduction channels between their open and closed states. The gating forces produced by channels' opening or closure result in an effective softening of the hair bundle. It's like pulling on a door with a rubber band: if the door abruptly opens, tension in the rubber band drops and the rubber band in turn feels softer. This phenomenon is called "gating compliance". When strong enough, gating compliance can yield a region of negative stiffness in the force-displacement relation of the hair bundle. Starting from a position at which all the channels are closed (on the left), the mean external force F that must be applied to bring the hair bundle to position X is reduced, with respect to the force that one would apply if the channels remained closed, by an amount corresponding to the total mean gating force $NZP_o(X)$ that is produced by channel opening. The force-displacement relation can be written as: $F(X) = K_\infty X - NZP_o(X) + F_0$, where $N \simeq 50$ represents the total number of transduction elements in the hair bundle (one per tip link), F_0 is a constant force that ensures $F(0) = 0$ and $P_o(X)$ is the sigmoidal relation between the channels' open probability and the position X of the hair bundle.

the stiffness of the hair bundle to become negative! In this case, the opening of a few channels results in the collective opening of all the others. Positions of negative stiffness are unstable; the hair bundle can only reside at a position of positive stiffness.

Although softer hair bundles respond with larger amplitudes to external forces, the nonlinear gating compliance fostered by synchronous channel gating cannot, by itself, be used to extract work from the hair bundle. The phenomenon of gating compliance, even in the case of negative stiffness, can be described by a nonlinear passive system of springs and gates at thermal equilibrium. For any amplificatory process, an energy source is mandatory.

3.3 Adaptation: an active feedback mechanism on the open probability of the transduction channels

The hair bundle is endowed with molecular motors that set the tip links under tension (Fig. 7.6a). This active tension allows the hair cell to operate within the steep region of its sigmoidal current-displacement relation. This condition is necessary for small hair-bundle deflections to elicit significant receptor currents, and thus for the transduction process to be sensitive to weak stimuli. Chemical disruption of the tip links results in channel closure, indicating that the transduction channels are intrinsically

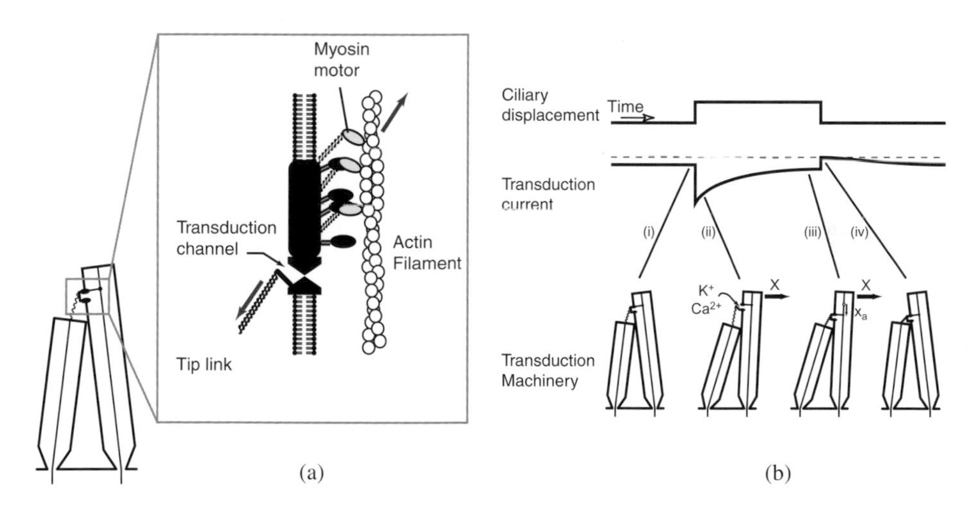

Figure 7.6. Adaptation motors. (a) Functional description of the transduction apparatus. A few dozens of myosin molecules anchor dynamically the transduction complex to the actin cytoskeleton of a stereocilium. At steady state, tip-link tension balances the force that is actively produced by the assembly of molecular motors towards the tip of the stereocilia. (b) Model of adaptation. An excitatory deflection X increases tip-link tension, opens the transduction channels, resulting in a receptor current. Although the position of the hair bundle is clamped at a fixed position, the magnitude of this current decreases as the result of an adaptive movement of the motors along the actin core of the stereocilia (here motion x_a towards the bottom) to restore the condition of force balance between the active motor force and tip-link tension. A deflection in the opposite direction would decrease tip-link tension, thereby eliciting an adaptive motor movement towards the tip of the stereocilia that would reset the tip-link under tension. Adapted from Hudspeth A.J. and Gillespie P., *Neuron.* 1994; 12:1–9.

more stable in a closed state. By pulling on the tip links, molecular motors ensure that a significant fraction of the channels are open at rest (Fig. 7.4e). Because each stereocilium is endowed with an actin core, it is not surprising that myosin motor molecules are involved in maintaining the tip links taut. Actin is indeed the natural substrate of active force production by myosin motors. Myosins exist in numerous isoforms and are involved in many cellular processes, including muscle contraction, intracellular protein transport, mitosis, cell motility, and morphogenesis. In the hair bundle, myosin motors exert an active force directed towards the tip of the stereocilia, a direction imposed by the polarity of the actin filaments. At steady state, this active force is balanced by an elastic restoring force in the tip link (Fig. 7.4a). Within this framework, the larger the active force that the motors can produce, the higher the open probability P_o of the transduction channels. For sensitive detection of hair-bundle movements, the active motor force must be tightly regulated to impose an open probability that is neither too small ($P_o \sim 0$) nor too large ($P_o \sim 1$). In frogs, steady-state tension has been estimated at 8 pN along the oblique tip-link axis. A few dozens of myosin molecules would suffice to sustain this tension. Because the hair bundle comprises ~50 transduction elements (one per stereocilium) that operate in parallel, the total active force that is produced by the motors is ~400 pN.

As with other sensory systems, the hair cell is endowed with an adaptation mechanism. Adaptation allows the cell to remain sensitive to weak sound stimuli even in the presence of a sustained deflection of the hair bundle that threatens to saturate mechano-electrical transduction. In response to a prolonged step displacement of the hair bundle in the excitatory direction, a hair cell generates a transduction current that first increases in magnitude, reflecting the opening of transduction channels, but then declines with time after reaching a peak because of channel reclosure. Conversely, channels reopen after they have been closed by an inhibitory stimulus. The molecular motors that are associated with the transduction apparatus are ideally suited to provide negative feedback on the open probability of the transduction channels (Fig. 7.6b). Starting from stall condition, at which the active motor force is precisely balanced by tip-link tension, the motors are expected to react to step deflections of the hair bundle by moving along the actin core of a stereocilium. By increasing tip-link tension, an excitatory stimulus promotes a slipping motion of the myosins that reduces tip-link tension, and in turn the channels' open probability, to their original values; similarly, a stimulus of opposite polarity allows climbing of the motor molecules towards the stereociliary tips, an active movement that rebuilds tension in the tip-links. In this model of adaptation, the molecular motors operate as active dynamical anchors of the transduction channel to the actin cytoskeleton of the stereocilia; they move to relax any stimulus-evoked change in tip-link tension.

If it were only to set the operating point of the transduction apparatus, adaptation could afford to operate on much slower timescales than the period of an acoustic stimulus. Yet, after being pulled open by an excitatory step stimulus, transduction channels reclose with typical timescales that are compatible with auditory frequencies. In addition, adaptation kinetics vary systematically along the tonotopic axis of auditory organs, with faster adaptation for cells devoted to the detection of high-frequency sounds. These observations suggest that adaptation provides high-pass filtering that may help in setting the hair cell's characteristic frequency of maximal responsiveness.

Myosin uses the energy derived from hydrolysis of adenosine tri-phosphate (ATP), which serves as biochemical fuel, to change conformation and set itself under tension. Transduction of chemical energy into mechanical work allows the motor to produce forces in the piconewton range. Adaptation, however, seems too fast to result solely from the ATP-ase activity of myosin molecules. Adaptation can indeed proceed over timescales that are shorter than $100\,\mu$s, whereas the ATP-ase cycle is three orders of magnitude longer ($\simeq$100 ms). To accelerate adaptation kinetics, the Ca^{2+} component of the transduction current provides fast feedback on the transduction channels' state to promote channel closure. Calcium increase results both in a decrease of the channels' open probability and in faster adaptation kinetics. Several mechanisms have been proposed to account for the effects of Ca^{2+} on adaptation. Calcium ions may bind directly to the transduction channel to stabilize its closed state. Alternatively, a protein associated to the channel may change conformation or reduce its stiffness to relax tip-link tension and in turn allow channel reclosure. In particular, Ca^{2+} ions may interact with the myosin motors that maintain the tip links under tension. By weakening the motors, i.e. decreasing the active motor force that pulls on the tip links, Ca^{2+} binding would result in channel closure. In any case, the binding site for Ca^{2+} must be found within only a few tens of nanometers of the channel's pore to account for the fast kinetics of adaptation. This condition would be hard to meet by myosins if, as a recent calcium-imaging study suggests, the transduction channels are located near the lower insertion point of the tip links (in contrast to the functional schematic shown in Fig. 7.6).

3.4 Active hair-bundle motility

Irrespective of the detailed mechanism for Ca^{2+} feedback, the reciprocal relation between channel gating and tip-link tension imposes that adaptive channel rearrangements evoke internal forces that drive active hair-bundle movements. Accordingly, various forms of mechanical excitability have been observed in response to force steps; they can all be understood as mechanical correlates of adaptation. In

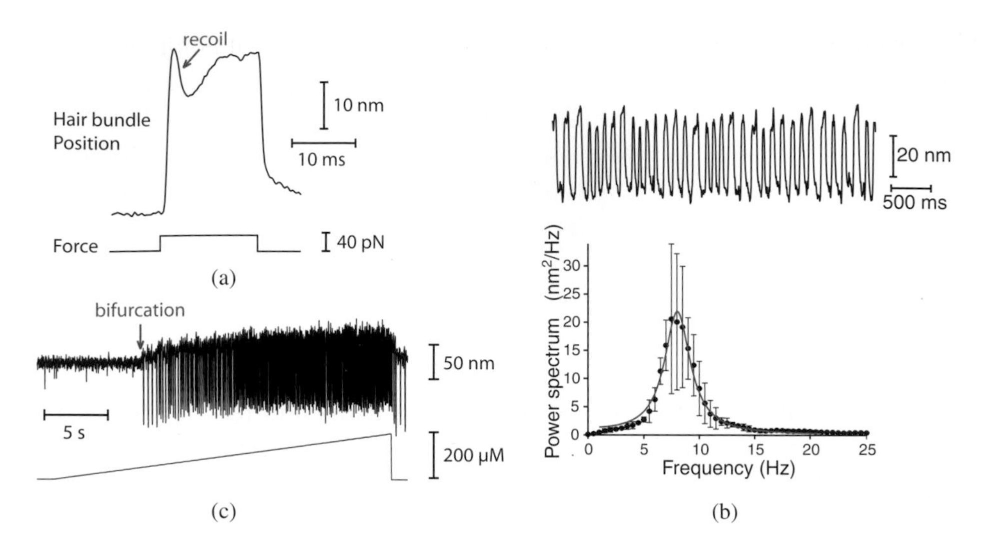

Figure 7.7. Active hair-bundle movements. (a) Twitch. A positive force applied to this quiescent hair bundle evokes a non-monotonic movement of the hair bundle. During the recoil, the hair bundle moves actively against the external force. (b) Spontaneous oscillation and its power spectrum. Note that the oscillations are noisy. (c) Increasing the calcium concentration above a threshold of a few tens of micromolars brings this quiescent hair bundle into a regime of spontaneous oscillations. (a) and (c) adapted from Tinevez J.Y. *et al.*, *Biophys. J.*, 2007; 93:4053–4067. (b) adapted from Martin P., Hudspeth A.J., Jülicher, F, *Proc. Natl. Acad. Sci. USA*, 2001; 98:14380–14385.

particular, when subjected to an abrupt step force in the excitatory direction, the hair bundle first moves in the direction of the applied force but can then display a transient recoil in the opposite direction, called the "twitch" (Fig. 7.7a). Twitches are mechanical analogs of action potentials in neurons. Strikingly, *in vitro* experiments in frogs, turtles, and eels have revealed that the hair bundle can also oscillate spontaneously (Fig. 7.7b). The power spectrum of bundle movement then displays a well-defined peak centered at the characteristic frequency of the spontaneous oscillation. Spontaneous oscillations, however, are noisy. Their regularity can be quantified by computing the ratio of the characteristic frequency and the width of the power spectrum at half the maximal height. This defines a quality factor $Q = 1-2$; the bundle movement loses phase coherence after a few cycles of oscillation.

Spontaneous hair-bundle oscillations are associated with the presence of negative stiffness in the bundle's force-displacement relation (Fig. 7.5b). Oscillations occur because Ca^{2+}-dependent adaptation continuously attempts to set the operating point of the transduction apparatus at an unstable position of negative stiffness. By affecting the adaptation process, calcium plays a key role as a control parameter of active hair-bundle motility. Oscillations occur only within a limited range of Ca^{2+} concentrations (0.1 – 1 mM in frog) and get faster in response to a mild

Ca^{2+} increase. In addition, large enough changes of Ca^{2+} elicit abrupt transitions between quiescent and oscillatory regimes of motility (Fig. 7.7c). These observations suggest that a hair bundle operates in the proximity of an oscillatory instability. In the physics of dynamical systems, this particular type of instability is called a Hopf bifurcation and will be discussed in more details later in this chapter.

3.5 The hair-bundle amplifier

As demonstrated by comparing responses to sinusoidal stimuli of oscillatory and quiescent hair bundles, spontaneous hair-bundle oscillations have important functional consequences on the mechano-sensitity of a hair cell (Fig. 7.8a). In the case of a quiescent hair bundle, the amplitude of the response is low and does not appear to depend on the stimulus frequency. In contrast, the response of an oscillatory hair bundle is tuned. Near the characteristic frequency of spontaneous oscillations, the bundle movement can be five-to-ten-fold larger than that of a quiescent hair bundle. The oscillations thus result in an active resonance of the hair bundle with the stimulus, and in turn, in an amplified response. For stimulation at resonance, the bundle deflection displays a compressive growth that contrasts with the linear response that is measured off-resonance (Fig. 7.8b). For intense stimuli, the response is about

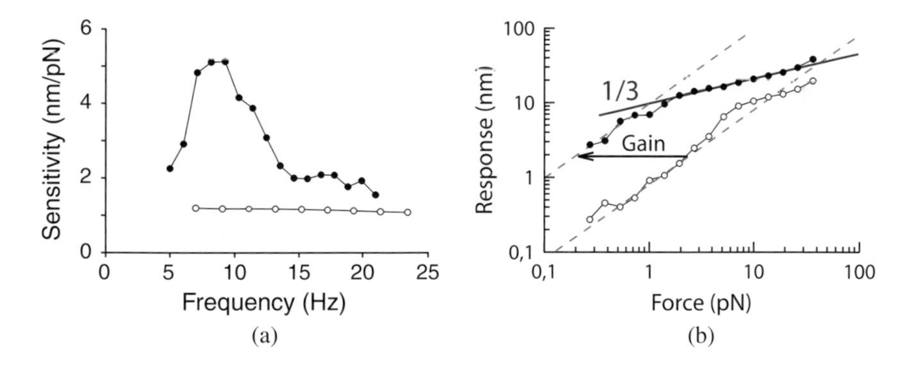

Figure 7.8. The hair-bundle amplifier. (a) A hair bundle that oscillates spontaneously at 8 Hz (black disks) displays resonance in response to a weak sinusoidal stimulus (here 4 pN) of varying frequency: the sensitivity of the hair bundle is maximal near the characteristic frequency of the spontaneous oscillation. A control cell whose hair bundle does not oscillate (white disks) instead shows a low sensitivity and no frequency selectivity. (b) For stimulation of an oscillatory hair bundle at resonance (black disks), its response X is amplified (gain: 5–10) at weak stimulus intensities F and evinces a compressive growth that is well described by a power law $X \propto F^{1/3}$. Off-resonance (white disks), here at a stimulus frequency ten times higher than the characteristic frequency of oscillations, the response is passive and growths linearly with the stimulus. Data in (a) from Martin P. and Hudspeth A.J., *Proc. Natl. Acad. Sci. USA*, 2001; 98:14386–14391 and in (b) from Jérémie Barral (unpublished; Laboratoire Physico-Chimie Curie, Paris, France).

the same at and off resonance and similar to that measured for a non-oscillatory cell; intense stimuli probe passive hair-bundle mechanics. The hair cell thus appears to amplify its response only when needed, for weak stimuli. Amplification by means of oscillations offers a double benefit for auditory detection. First, by amplifying weak stimuli, the oscillatory hair bundle lowers the stimulus intensity required to elicit a vibration large enough to be detected (on the order of 1 nm) and thus enlarges the dynamical range of hearing. Second, because amplification is effective only near the characteristic frequency of oscillation, the oscillator sharpens frequency selectivity by filtering the input to the hair cell. We note, however, that amplification by means of oscillation results in a conspicuous compressive nonlinearity that ought to distort the input to the hair cell. The hair-bundle amplifier is thus not a high fidelity sound receiver. We have seen earlier that our perception of sound also does not faithfully reflect the acoustic input. The nonlinear behavior of the hair-cell bundle may explain why.

A careful reader will have noticed that we find here, at the level of a single hair cell (from a frog!), a behavior that resembles that of the cochlear amplifier, which, at a very different scale, enhances the sensitivity and frequency selectivity of mammalian hearing (Fig. 7.3). This similarity qualifies the hair-bundle amplifier as a plausible component of the active process that underlies the remarkable properties of the auditory system. As a matter of fact, in non-mammalian vertebrates, the hair-bundle amplifier is the only candidate mechanism to explain auditory amplification. This doesn't imply, however, that the hair bundle is the only force generator in the ear.

3.6 Somatic electromotility of cochlear outer hair cells in the mammalian cochlea

In mammals, the cochlea contains two types of hair cells (Fig. 7.9a). Inner hair cells are the true sensory cells. Abundantly innervated by afferent fibers, they send to the brain an electrical signal that reports transverse mechanical vibrations of the cochlear partition. Outer hair cells, which represent three quarters of all cochlear hair cells (i.e. $\simeq$9,000 cells), send almost no information to the brain! Instead, they receive an efferent innervation which, when activated to mimic a signal coming from the brain, reduces the mechanical sensitivity of the partition to weak sound stimuli. In addition, outer hair cells display a unique form of motility called electromotility. Electromotility is a biological analog of piezoelectricity, by which the soma of the outer hair cell contracts or elongates in response to depolarization or hyperpolarization of the transmembrane potential, respectively, with a sensitivity of 20 μm/V (Fig. 7.9b). This process is based on a charge movement (possibly Cl^-) through prestin, a transmembrane protein that belongs to a family of anion transporters and

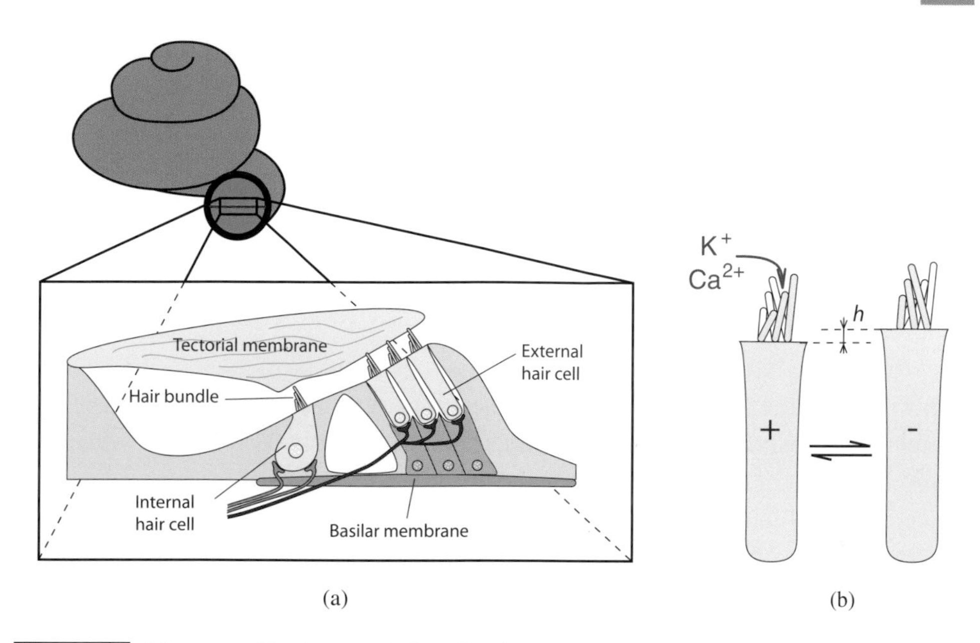

Figure 7.9. Electromotility by outer hair cells of the mammalian cochlea. (a) The organ of Corti. The cochlear partition, here shown in cross-section, contains two types of hair cells. "Inner" hair cells receive abundant afferent innervation (green; signal from the cell to the brain), whereas the "outer" hair cells are almost exclusively innervated by efferent fibers (red; signal from the brain to the cell). The sound-pressure wave evokes a pressure difference across the partition that sets the basilar membrane into vibration. This transverse vibration evokes a shearing motion of the apical surface of the hair cells relative to the tectorial membrane, which deflects the hair bundles. (b) Electromotility. The soma of outer hair cells change length as a function of the transmembrane electrical potential. An excitatory deflection of the hair bundle (left) evokes cationic influx that depolarizes the membrane (+), which results in the shortening of the cell soma. In the other half period of the auditory stimulus (right), the cell membrane is hyperpolarized (−) and the soma gets longer. The amplitude of this movement, here noted h, can reach one micrometer for saturating stimuli.

is inserted at high density (4 000–6 000 μm^{-2}) in the somatic membrane of the cell. By changing the surface that it occupies in the membrane, prestin changes the length of the cell. Under isometric conditions, the force produced by this process is 50 pN/mV, which yields a maximal force per cell on the order of one nanonewton. Genetic and acute inactivation of prestin have demonstrated that electromotility is a necessary component of the cochlear amplifier.

Electromotility, however, is a linear process: for physiological stimuli, somatic length changes increase linearly with the membrane potential. Moreover, this mechanism depends only weakly on the stimulus frequency. Thus, it cannot, by itself, account for cochlear amplification. The active process is indeed characterized by an essential compressive nonlinearity that depends very heavily on frequency (Fig. 7.3). It is likely that amplification results from a complex interplay between the two forms of motility that we have discussed here: hair-bundle motility and electromotility. Their detailed interaction, however, has not been clarified and is currently the subject of intense research.

4 The "critical" oscillator as a general principle of auditory detection

As already noted, the hair-bundle oscillator, which has been studied *in vitro* at the level of a single hair cell, recapitulates the four cardinal properties of the cochlear amplifier. First, the stimulus is amplified: the intensity that evokes a vibration beyond a detectable threshold is lowered. Second, frequency selectivity is sharpened: amplification is efficient only near a characteristic frequency, which results in an active resonance that filters the input to the sensory system. Third, for stimulation near the characteristic frequency, and only there, the response displays a compressive nonlinearity: a wide range of stimulus intensities is represented by a much narrower range of vibration amplitudes. This nonlinearity may explain the "phantom tones" that are perceived in response to two-tone stimuli. Finally, the mechanical activity can produce, even in the absence of a stimulus, self-sustained oscillations that could account for oto-acoustic emissions.

These four properties have been recognized as signatures of a dynamical system that operates close to an oscillatory instability, termed a Hopf bifurcation. A Hopf bifurcation occurs when there is an abrupt transition from quiescence to spontaneous oscillation upon continuous variation of a quantity describing some component of the system, which serves as "control parameter". We have seen that in the case of the hair-cell bundle, the calcium concentration plays the role of a control parameter (Fig. 7.7c). For such an active system, there is a critical value of the control parameter at which oscillations appear or vanish. For this reason, a dynamical system that is poised at a Hopf bifurcation is called a "critical oscillator". The response of a critical oscillator to sinusoidal stimuli near the characteristic frequency of the oscillator displays generic properties, i.e. properties that do not depend on the details of the active mechanism that results in the oscillatory behavior (see the text Box 7.1). These properties are characteristic both of a single hair-bundle oscillator and of the cochlear amplifier. In particular, vibrations display a compressive growth that follows a power law 1/3 (Fig. 7.10b). A critical oscillator is ideally suited for auditory detection, even if its inherent nonlinearity yields pronounced distortions and masking in response to complex sound stimuli. Two-tone interferences in fact appear as a necessary price to pay for the exquisite sensitivity, the sharp frequency selectivity, and the wide dynamical range of auditory detection by a critical oscillator.

Box 7.1. Hopf bifurcation: behavior of a critical oscillator.

Many dynamical systems are able to oscillate spontaneously, including lasers, chemical reactions, neuronal networks, or biological cells such as the hair cells that are the subject of the present chapter. Being governed by many coupled degrees of freedom, the behavior of such systems is generally complex. However, the description of a dynamical system can be greatly simplified if the system operates close to a Hopf bifurcation. Near the oscillatory instability, only two degrees of freedom suffice to describe, at long times, the relaxation dynamics of the system to an external perturbation as well as its response to sinusoidal stimuli at frequencies near the characteristic frequency $f_C = \omega_C/(2\pi)$ of the instability. In the case of the hair-cell bundle, these two degrees of freedom have been identified. They are the position X of the hair bundle and x_a of the molecular motors that set the tip links under tension (Fig. 7.6). In addition, through an analytic, but nonlinear, sequence of variable changes, one can describe the dynamical coupling between these two variables by a single equation — called the "normal form " — of a complex variable Y:

$$\frac{dY}{dt} = -(\mu_C - \mu - i\omega_C)Y - b|Y|^2 Y + \frac{F}{\Lambda}, \tag{7.1}$$

where $X \equiv \text{Re}(Y)$ represents the position of the system, Λ a friction coefficient and F the external force. To apprehend the behavior of this equation, let's forget for a moment the nonlinear term of amplitude b and the external force. As long as the control parameter μ is smaller than the critical value μ_C, a perturbation imposing an initial condition $Y = Y_0 \neq 0$ evokes a damped oscillation towards the stable value $Y = 0$. In contrast, if $\mu > \mu_C$, we get an oscillation with an amplitude that grows exponentially. When the oscillation amplitude gets too large, the nonlinear cubic term asserts itself to prevent divergence of the movement and stabilizes the amplitude of motion at a finite value (we need for that to impose $\text{Re}(b) > 0$). At the critical value μ_C of the control parameter μ, the system displays a transition between a quiescent state and an oscillatory state: it's the Hopf bifurcation.

Let us now assume that the system is stable ($\mu < \mu_C$) and that it experiences a sinusoidal force $F = \bar{F}e^{i\omega t}$ (here written using complex notations) of angular frequency ω. Then, we immediately get from the normal form that the response $Y = \bar{X}e^{i\omega t}$ at the frequency of stimulation must obey:

$$\bar{F} = A\bar{X} + B|\bar{X}|^2\bar{X}, \tag{7.2}$$

where the impedance $A = \Lambda[(\mu_C - \mu) + i(\omega - \omega_C)]$ and the term $B = \Lambda b$ describe the linear and nonlinear components of the response, respectively. In the regime of weak forces, the system show a sensitivity $\chi = |\bar{X}|/|\bar{F}| = 1/|A|$. We thus get a resonance with sensitivity that peaks at the characteristic frequency $\omega = \omega_C$ of the oscillator (Fig. 7.10a). The width of the resonance is given by $\Delta\omega_{\text{lin}} \cong 1/|\mu_C - \mu|$.

At the bifurcation $\mu = \mu_C$, the system displays a surprising feature: when stimulated at resonance ($\omega = \omega_C$), the impedance A vanishes. The response of the system is then under the control of the cubic nonlinear term and displays a compressive growth that follows a power law: $\bar{X} \propto \bar{F}^{1/3}$ (Fig. 7.10b). The sensitivity χ in turn varies as $F^{-2/3}$ and thus formally diverges when the force goes to zero. Because there is no force weak enough to elicit a linear response, one talks about an "essential" nonlinearity to describe the system's behavior near the critical point. Note that this kind of behavior can only be produced by an active system. A passive system cannot cancel the hydrodynamic friction that impedes the periodic movement evoked by the external force. For this to happen, the system must be capable of mobilizing energy resources to provide mechanical work that produces "negative friction".

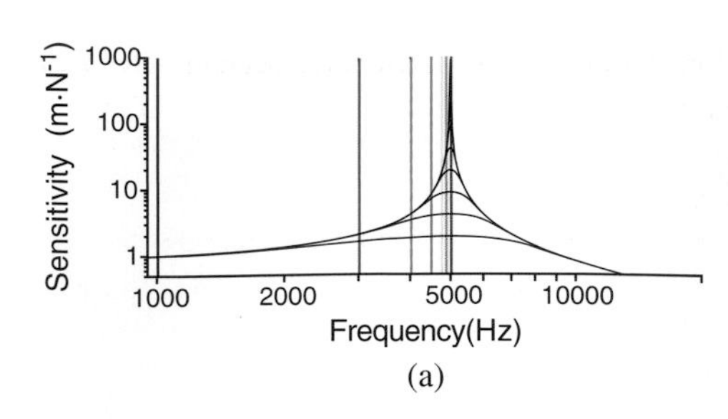

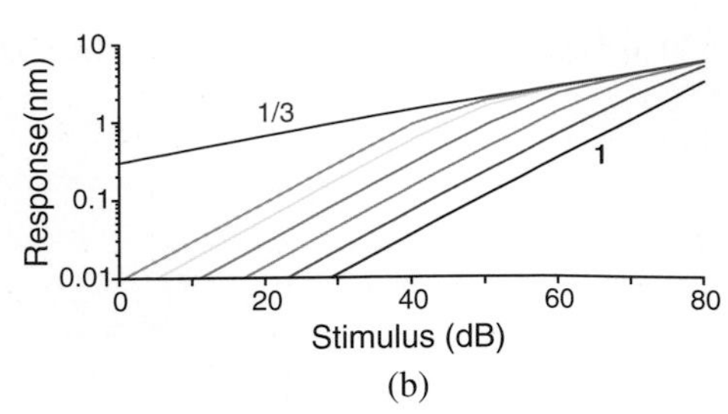

Figure 7.10. Amplification by critical oscillation ($f_C = 5$ kHz; $\mu = \mu_C$). (a) Sensitivity $\chi = |\bar{X}|/|\bar{F}|$ to a sinusoidal stimulus $F = \bar{F}e^{i\omega t}$ as a function of stimulus frequency $\omega/(2\pi)$ for six intensities $\bar{F}$ of stimulation (from top to bottom: 0, 40, 50, 60, 70, 80 dB). (b) Amplitude of the response $\bar{X}$ as a function of the intensity $\bar{F}$ of the stimulus for the 7 frequencies marked in A by colored lines. (Figure adapted from Hudspeth A.J., Jülicher F., Martin P., *J. Neurophysiol.* 2010; 104:1219–29.)

Although the detailed subcellular and molecular mechanisms that allows the hair cells to amplify their input is not fully established, it seems clear that auditory detection benefits from the activity of self-sustained oscillators in the ear. A single oscillator is a good sound detector only near its characteristic frequency of spontaneous oscillations. The analysis of complex sound stimuli such as speech or music thus calls for the operation of an assembly of oscillators with distributed characteristic frequencies in the auditory range (20 Hz to 20 kHz in human hearing).

Bibliography

[1] http://www.neuroreille.com/promenade/english/start_gb.htm. For a promenade around the cochlea with many illustrations and animations.

[2] Moore BCJ. *An introduction to the psychology of hearing*, Elsevier (2004). An introduction to human psychoacoustics.

[3] Goldstein JL. Auditory nonlinearity. *J. Acoust. Soc. Am.* (1967) **41**, 676–89. A pionneer study on the auditory nonlinearity from the characterization of *phantom tones* that are perceived by the human ear in response to two-tone stimuli.

[4] Manley GA. Evidence for an active process and a cochlear amplifier in nonmammals. *J. Neurophysiol.* (2001) **86**, 541–9. This study compares the auditory specifications of various species.

[5] Robles L, Ruggero MA. Mechanics of the mammalian cochlea. *Physiol. Rev.* (2001) **81**, 1305–52. A detailed review of the data on cochlear mechanics that led to the emergence of the concept of cochlear amplification.

[6] Hudspeth AJ. Making an effort to listen: mechanical amplification in the ear. *Neuron* (2008) **59**, 530–45. A review article that makes the connection between auditory amplification and active motility by the mechanosensory hair cells in the inner ear.

[7] Martin P. Active hair-bundle motility of the hair cells of vestibular and auditory organs. *Active Processes and Otoacoustic Emissions in Hearing* (Manley GA, Popper AN, Fay RR, eds), New York,

Springer (2008), 93–143. A complete description of the hair bundle: structure, mechano-electrical transduction, passive and active mechanical properties, oscillations, and their modelisation.

[8] Ashmore J. Cochlear outer hair cell motility. *Physiol. Rev.* (2008) **88**, 173–210. A detailed review on somatic electromotility of outer hair cells in the mammalian cochlea.

[9] http://www.ucl.ac.uk/silva/ear/research/ashmorelab. A spectacular video that illustrates the phenomenon of electromotility: an outer hair cell, isolated from the cochlea of a guinea pig, dances in response to rock music!

8

Sensing through friction: the biomechanics of texture perception in rodents and primates

Georges Debrégeas, *CNRS, Laboratoire Jean Perrin, UPMC, Paris.*
Yves Boubenec, *CNRS, Unité de Neurosciences, Information et Complexité, Gif-sur-Yvette. Present address: Institut d'Etude de la Cognition, Equipe Audition, ENS, Paris.*

Rodents and primates possess an exquisite tactile sensitivity, which allows them to extract a wealth of information about their immediate environment. They can distinguish subtle differences in surface roughness through tactile exploration in a much more precise way than they can do visually. In both sensory systems, tactile information is contained in the sequence of deformation of the tactile organ — the facial hair for rodents (the whiskers), the digital skin for primates — elicited by active rubbing on the probed surface (Figure 8.1). These deformations, registered by mechanosensitive neurons located in inner tissues, are processed by the central nervous system to produce a sensory representation of the surface.

In both sensory systems, the biomechanics of the tactile organ plays an important role. It controls the way surface geometrical properties are converted *via* frictional interaction into mechanical energy that triggers the neural response. This mechanical transduction process effectively shapes the sensory signal, and thus participates in both the filtering and coding of tactile information. Just as the comprehension of the

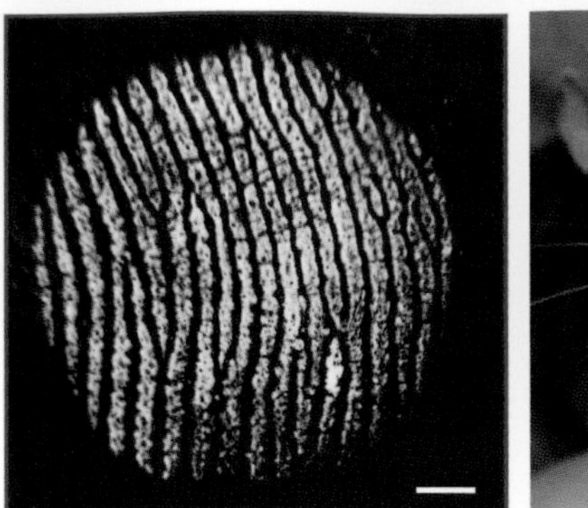

Figure 8.1. Tactile perception in primates and rodents. Left: in primates, tactile information is conveyed through the digital skin. Right: rodents preferentially use their long facial hair, or whiskers, to explore their environment.

eye's optical properties is necessary to understand vision, fully apprehending tactile perception requires understanding this mechanical process. The striking differences between the mechanical and geometrical properties of the whisker and the skin suggest that different encoding strategies may be at play in each of them.

1 Cutaneous mechanoreceptors of the human glabrous skin

Skin acts as a protective barrier from physical, microbiological and chemical attacks. It is also the most extended sensory organ of the human body. At the adult age, its surface is $\approx 2\,m^2$, compared to a few mm^2 for the retina. Tactile sensitivity greatly varies across the body. Spatial resolution for instance, i.e. the minimum distance between two localized stimuli that can be distinguished, ranges from 2 to 40 mm. Tactile accuracy is maximum in the glabrous part of the skin, notably the hands and lips, reflecting the crucial role of tactile perception for motor control during manipulation tasks.

Cutaneous sensitivity relies on populations of specialized nerve endings that innervate the skin. Each afferent fiber is associated with one (or several) mechanosensitive endings located 0.5 to 2 mm below the skin surface. Their density is maximum in the glabrous skin, and in particular in the hand (17,000 units per hand). Mechanoreceptors convert mechanical energy, produced by the skin deformation, into electrical signals. The transduction process relies on the presence of mechanosensitive ionic channels in the membrane of mechanoreceptors, which modify the ionic permeability in response to a change in the membrane geometrical

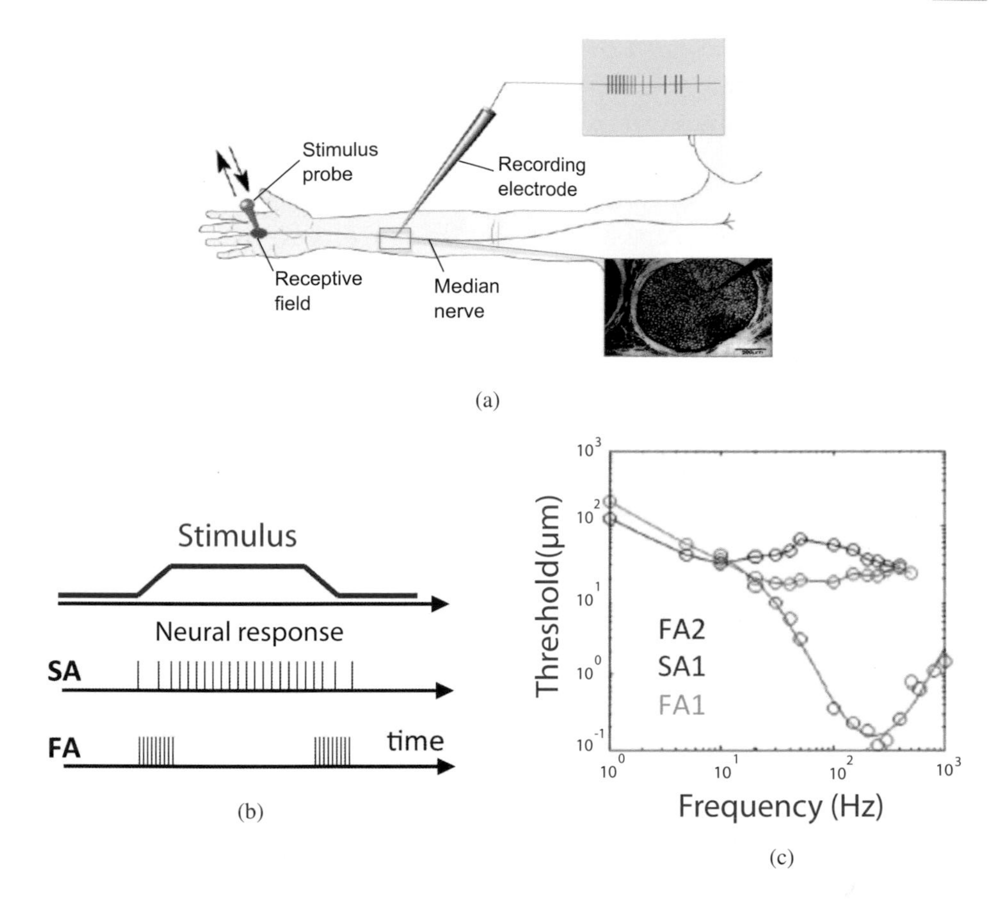

Figure 8.2. Neurophysiological characterization of the mechanoreceptive fibers. (a) Experimental scheme for neural recordings in the human median nerve. The blow up shows an image of the recording electrode within the bundle of tactile fibers. (b) FA and SA spiking responses to a ramp and hold stimulation. (c) Frequency-dependent response threshold of SA1, FA1 and FA2 fibers (adapted from Muniak *et al.*, *J. Neurosci.* (2007) **27**(43), 11687–11699).

conformation. Above a certain threshold, the induced depolarisation triggers an action potential that propagates along the axon to the central nervous system.

In order to probe the functional characteristics of mechanoreceptors in human subjects, neurophysiologists introduce a thin electrode in the median nerve (Fig. 8.2a). They can record the spiking sequence of individual afferent fibers in response to controlled skin stimulations. These experiments allowed them to identify four distinct populations of mechanosensitive fibers that innervate the glabrous skin. They are classified according to two characteristic features of their neural response to a localized indentation of the skin. The first trait is the receptive field, defined as the area of the skin over which a given intensity of stimulation produces a measurable response. Class I receptors exhibit small and well-defined receptive fields, while class II receptors are associated with large receptive fields with fuzzy edges. The second characteristic is the response adaptation to a sustained stimulus. A ramp and hold indentation is applied in the receptive field (Fig. 8.2b). Slowly adapting receptors (SA) continuously discharge during the whole stimulation phase while rapidly adapting receptors (FA) respond to the time-varying periods, but remain silent when the stimulation is maintained at a constant level. To quantitatively characterize the temporal aspect of the response, a sinusoidally oscillating stimulation is applied. For each frequency, the threshold amplitude that triggers a measurable response is extracted (Fig. 8.2c). For SA receptors, the sensitivity is found to be rather independent of the oscillatory frequency, while FA receptors exhibit an optimum sensitivity within a well defined frequency range.

The distinct response properties of the four classes of fibers reflect differences, in physiology and locations within the skin, of their associated mechanosensitive endings (Figure 8.3). SA1 mechanoreceptors terminate in Merkel's cell complexes, which consist in highly branched structures located at the junction between the derma and epiderma. SA2 receptors' terminations, the Ruffini's endings, have the shape of $\approx$1 mm elongated ellipsoids and respond principally to stretching along their major axis. FA1 fibers terminate in Meissner corpuscles, located in the dermal papillae, these regular inter-digitations of the derma. Finally, the FA2 fibers terminate in Pacinian corpuscles (PCs), which are large onion-like structures embedded in the deeper layers of the derma. Besides their various physiological characteristics, the different classes of mechanoreceptors also vary in the density of innervation. Box 8.1 sums up the essential properties of each mechanoreceptive unit.

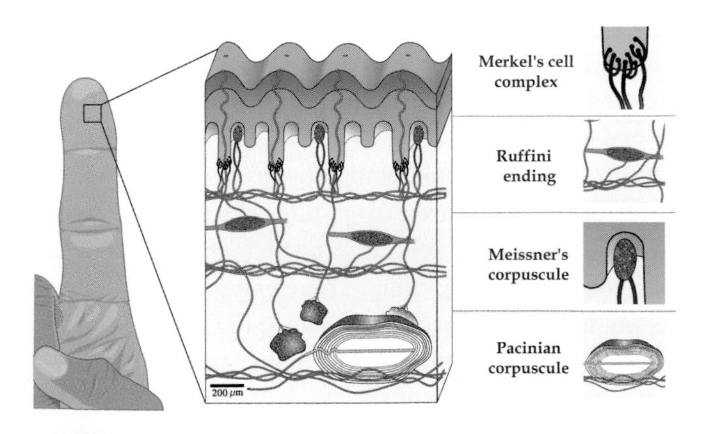

Figure 8.3. The four mechanoreceptors of the human glabrous skin.

Box 8.1 Characteristics of the cutaneous mechanoreceptors.

Receptor	SA1	SA2	FA1	FA2
nerve ending	Merkel's cell	Ruffini ending	Meissner's corpuscle	Pacinian corpuscle
depth (mm)	0.5	0.3	1	2
density (unit/cm^2)	70	30	130	15
Receptive field (mm^2)	10	10	60	>100
frequency tuning (f_{opt})	weak	weak	strong (50 Hz)	strong (200 Hz)

2 The neural basis of roughness perception

Owing to these various response characteristics, the four classes of mechanoreceptors are bound to convey different types of information, in the same way as cone and rod cells in the retina carry different features of the image impinging on the retina. The perception of roughness, i.e. the geometrical texturation of a surface, illustrates this parallel processing scheme.

Trained humans are able to read Braille, by feeling the particular arrangement of sub-millimeter raised dots, but they can also differentiate between essences of woods or quality of fabrics. In the late 1920's, on the sole basis of psychophysical assays, David Katz suggested that roughness perception may involve two independent coding channels specific to the perception of coarse and fine textures. He noticed that the judgment of coarse textures (associated with features of lateral dimensions larger than about 200 μm) could be readily made by statically pressing the skin on the surface, without lateral movement. This suggested that spatial cues (differences in induced skin deformation across the contact) are central to such roughness evaluation. In contrast, when evaluating finely roughened substrates (such as paper or fabrics), the absence of relative skin motion greatly impairs texture perception. This in turn points towards vibratory encoding of fine roughness: discrimination would here rely on differences in the spectral content of the friction-induced skin vibrations, which would constitute the equivalent of a tactile color. In the last two decades, this so-called *duplex theory* of roughness perception has received confirmation through a series of experiments that combined psychophysical assays and neurophysiological recordings of afferent fibers in both humans and monkeys. These experiments further clarified the categories of mechanoreceptors that mediate coarse and fine roughness perception, respectively.

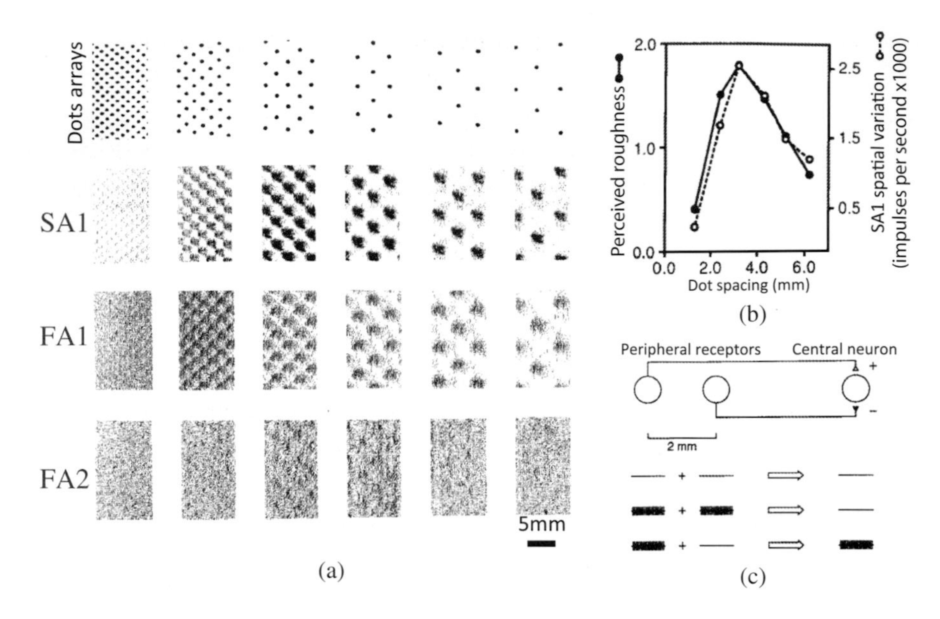

Figure 8.4. Neural responses to raised dots. (a) Top: stimulus used. Each circle represents a raised dot of diameter 0.5 mm. In the three following lines, each dot represent one action potential. The 2D map is obtained by scanning the texture along parallel lines shifted vertically by 0.2 mm (adapted from Connor *et al.*, *J. Neurosci.* (1990) **10**(2), 382–383). (b) Comparison between the roughness magnitude estimate from psychophysical experiments and the spatial variations of SA1 firing rate computed from neurophysiological recordings. (c) Spatial mechanism for roughness perception: a central neuron, that receives excitatory (+) and inhibitory (−) inputs from two neighboring SA1 receptors (separated by 2 mm), will fire only when one out of the two receptors is activated, which corresponds to textures with dot spacing ≈ 4 mm (adapted from Connor and Johnson, *J. Neurosci.*, 1992, **12**(9):3414–3426).

In the early 1990s, Connor and Johnson ran a series of experiments to investigate the neural basis of coarse roughness perception. They used surfaces exhibiting regular arrays of elevated dots of various sizes and spacings as stimulation substrates. They recorded mechanoreceptors' discharge rate (impulses per second) in monkeys as such substrates were swept across their receptive fields (Fig. 8.4a). By continuously shifting the stimulus array relative to the mechanoreceptor, they were able to produce a detailed spatial map of each population of receptor response. The same type of substrates were further used in psychophysical experiments: participants were asked to assess the perceived roughness evoked by the stimuli on a one to ten scale. The latter experiment established that perceived roughness magnitude is a nonmonotonic function of dot-spacing: it increases with dot-spacing up to 3.2 mm then declines. Hypothetical neural codes were then computed from the mechanoreceptor's response, and evaluated based on their ability to account for the psychophysical result. Intensive codes, relying on average discharge rates, were shown to be inconsistent. The most plausible code relied on spatial variation of the firing rate between neighboring SA1 receptors. This spatial differentiation processing was suggested to result from the integration of central neurons of excitatory and inhibitory inputs from two mechanosensitive units with close-by receptive fields, as illustrated in Fig. 8.4c. Such cells are expected to exhibit a maximum response upon activation of one receptor while the other remains silent.

The neural basis of fine texture perception is far less established. The fact that fine texture discrimination relies on FA2 (PC) receptors was however clearly demonstrated using adaptation experiments. By applying vibrations of moderate intensity to the skin for a few minutes, it is possible to saturate FA channels and momentarily reduce their sensitivity. Such pre-stimulation was found to substantially impair tactile discrimination of finely roughened surfaces while leaving coarse texture judgement unchanged. This effect was found to be significant only when the adaptation vibration frequency fell within the range of the Pacinian channel (200–300 Hz), which indicated that PCs played a more central role in such tasks.

3 Mechanical filtering of tactile information by the skin tissue

Mechanoreceptors are local mechanical sensors: their response is controlled by a particular combination of the stress and/or strain imposed on the mechanosensitive ending site. This *relevant stimulus* is expected to be specific to each population of mechanoreceptors, but its exact nature still remains largely unknown: although it is possible to record the neural signal from individual afferent fibers elicited by a given stimulation of the skin surface in both humans and monkeys, there is currently no method to simultaneously monitor the associated local mechanical state at the mechanoreceptor site. Fully understanding tactile sensing thus requires the modeling of the skin as a biomechanical interface in order to infer how the surface stress field, induced by a contact between the skin and the touched object, propagates down to the subcutaneous receptors.

Several biomechanical models have been proposed to describe the skin transduction mechanism. The simplest approach consists in considering the skin and subcutaneous tissues as a homogeneous elastic medium. Under this assumption, for any component of the stress or the strain (or a linear combination of them), one may compute the associated field $G(x, z)$ that results from the application of a localized unit force applied at position ($x = 0$, $z = 0$) at the surface the skin (Fig. 8.5a) — for the sake of simplicity, we will use a 2D biomechanical description, assuming invariance along the y direction. For a given surface pressure profile $\sigma(x)$,

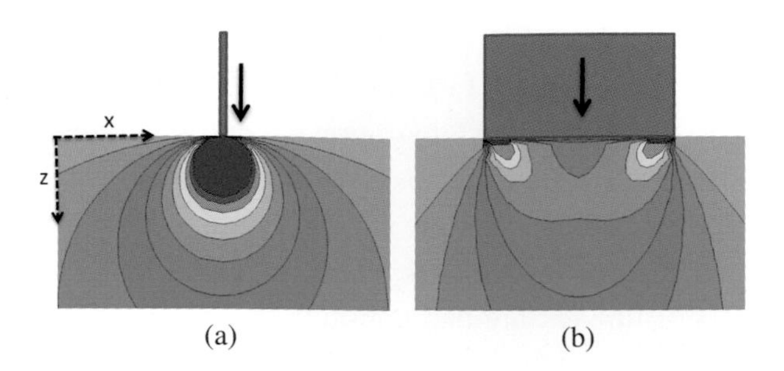

Figure 8.5. Compressive vertical stress field inside an elastic bloc submitted to a localized force (a) and a rectangular indentor (b).

the relevant subcutaneous profile experienced by a population of mechanoreceptors located at a depth h reads:

$$p(x) = \int G(x - x', h)\, \sigma(x')\, dx' \tag{8.1}$$

The signal effectively "seen" by the mechanoreceptors is thus a (discretized) blurred version of the superficial stimulation. The filtering process operated by the skin tissue is entirely characterized by the function $F(x) = G(-x, h)$, which can be seen as the biomechanical receptive field of the receptor.

The spatial shape of the filtering function F depends on the relevant mechanical stimulus that controls the mechanoreceptor's response. The form of the elastic equations however results in a unique scaling property such that $G(x, h) = 1/h\, G(x/h, 1)$. This relationship indicates that the characteristic width of $F(x)$ increases linearly with the depth h at which the mechanoreceptor is embedded within the skin. This simple mechanical analysis thus provides a direct interpretation of the observation that mechanoreceptors embedded in the deepest layers of the derma exhibit more extended receptive fields. As a rule of thumb, receptors laying at a depth h are expected to be insensitive to surface features of dimensions smaller than $\approx h$.

Another important consequence of the the skin mechanical transduction can be deduced from the analysis of the indentation by a rectangular object. The resulting pressure field displays strong divergences at the edge of the indentor (Fig. 8.5b). This stress focusing effect provides an important nonlinear mechanism of contrast reinforcement for the tactile system by enhancing the effective stimulation signal along the edges of a topography.

3.1 Mechanical transduction of surface topography under static load

This biomechanical model was initially developed by Johnson and Philips in order to obtain a comprehensive description of the response of SA1 populations to indented spatial patterns. The discharge rate (number of spikes per second) of individual SA1

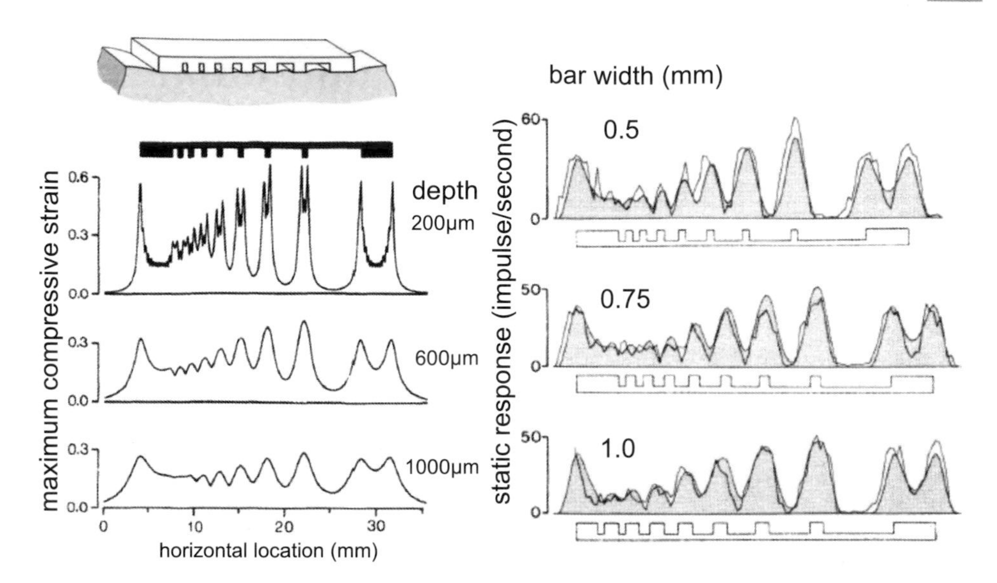

Figure 8.6. SA1 fibers response to gratings. Scheme of the indenter pressed onto a semi elastic bloc. Left: maximum compressive strain computed at different depth within the skin. Right: comparison between the maximum strain profiles calculated at a depth of 500 μm (grey) and the SA1 discharge rate (thin line, measured in impulse per second) for three different gratings. The SA1 response profiles were obtained by moving the grating step-wise across the fiber receptive field. Adapted from Johnson, *J. Neurophysi.* (1981) **46**(6) 1192–1203.

fibers innervating the monkey fingerpad was measured while indenting the skin with various rectangular gratings under steady load (see Fig. 8.6). For each grating, a receptor's response profile was obtained by sweeping the pattern stepwise across the fingerpad. The biomechanical modeling consists of two steps. First, the surface pressure profile $\sigma(x)$ produced by the indentor is numerically computed. Various strain and stress fields are then calculated at different depth h based on Eq. (8.1). These different filtered signals are then confronted with the measured SA1 response profiles in order to identify which component of the local tissue deformation is most predictive of the neural response. This method allowed the authors to establish that the maximum compressive strain calculated at a depth of 500 μm (consistent with the position of Merkel's cells within the skin) constituted the best candidate for the relevant mechanical stimulus of SA1 receptors, as it provides the best match with the recorded spike rates for a large collection of gratings.

3.2 Texture transduction under active exploration

The preceding section addressed a configuration in which the stimulation was statically applied to the fingerpad. The same approach may be implemented to compute the temporal signal elicited at the mechanoreceptive site during a dynamic exploration. When the substrate is moved across the finger at velocity v along the direction x, the pressure profile on the skin surface is continuously translated and can be written as $\sigma(x-u)$ where $u = vt$ is the instantaneous substrate position. The mechanical signal driving the response of a mechanoreceptor located at $x = 0$ then reads $p(u = vt) = \int F(x) \cdot \sigma(x-u)dx$. Each spatial modulation of the surface pressure of wavelength l thus produces a modulation of the local signal at frequency v/l. The amplitude of the oscillating signal is given by the product of the Fourier transform of F and σ at this particular spatial mode. Since F decays over length of order h, its spectrum has the characteristic of a low-pass filter with a cut-off frequency $1/h$. Signals associated with surface corrugations of lengthscale below h are thus expected to have vanishingly small amplitudes.

This observation tends to indicate that FA2 receptors should be insensitive to textures of characteristic lengthscales smaller than a few mm. This is at odds with the finding that these deepest receptors principally mediate the tactile perception of fine features ($<200\ \mu$m). This apparent paradox suggests the existence of other mechanisms that would facilitate the detection of friction-induced cutaneous vibrations.

3.3 Spectral selection by the fingerprints

The preceding description implicitly assumed that, as a texture is swept across the fingertip, the contact pressure distribution gets similarly translated. This approach ignores the fact that the contact region between the fingertip and the surface has a finite area. If the surface is smooth, the pressure distribution across the contact, noted $\bar{\sigma}$, is not uniform: it is maximum at the center and decays to zero at the border (see Fig. 8.7a). Ignoring edge effects, one may approximate the stress profile induced by a fine grating as $\bar{\sigma}(x)T(x)$, where T is the topography of the substrate. This expression reflects the fact that the envelope of the surface stress is set by the large-scale geometry of the fingertip. The subcutaneous signal upon active exploration thus becomes $p(u) = \int (F \cdot \bar{\sigma})(x) \cdot T(x-u)dx$. The transformation of the topographical (not the pressure) profile into a subcutaneous signal is thus controled by the function $g = F \cdot \bar{\sigma}$. In this product, the function F contains the intrinsic properties of the receptor within the skin, while the reference field $\bar{\sigma}$ depends on the context of the tactile exploration, a set of conditions that can be (in part) tuned by the individual. It includes the force applied by the fingertip onto the surface, the skin/substrate friction coefficient, or the position of the contact zone with respect to the sensor location. This pressure field can be seen as the analog of the illumination field in the visual system, as it controls the way the mechanical energy is distributed across the skin surface and hence how different regions of the substrate contribute to the overall signal.

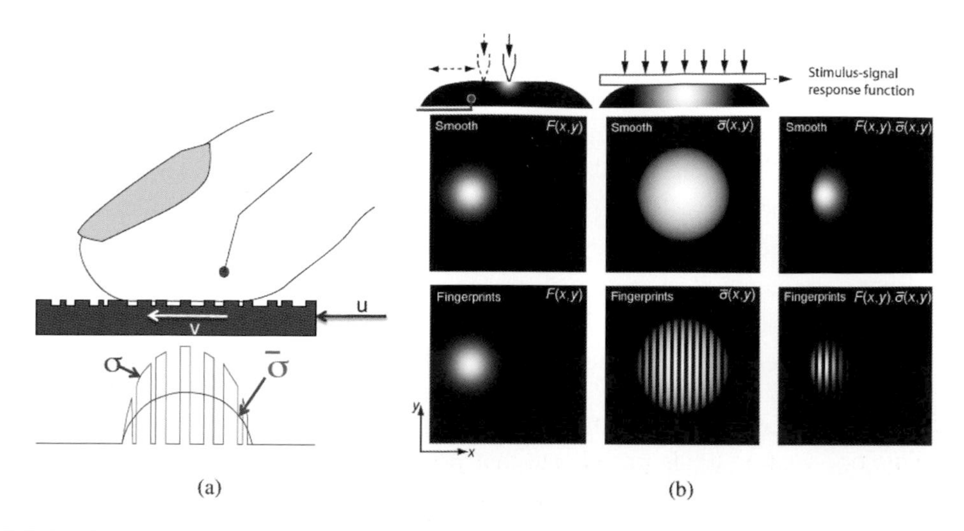

Figure 8.7. Effect of fingerprints on mechanical transduction of fine textures. (a) Instantaneous pressure profile $\sigma(x)$ induced at the surface of the skin by a grating scanned at velocity v. The envelope of the profile $\bar{\sigma}(x)$ (in red) corresponds to the smooth situation. (b) Comparison of the filtering function g for a smooth and fingerprinted skin. g is the product of the impulse response function F and of the reference field $\bar{\sigma}$. Unlike F, the pressure field $\bar{\sigma}$ is strongly modified by the presence of fingerprints. The resulting filtering function g displays large spatial modulations at the wavelength of the fingerprints.

At first sight, the reference field might seem irrelevant, since its characteristic extension is in the centimetric range, such that $F \cdot \bar{\sigma}$ should be rather similar to F itself. However, the presence of epidermal ridges (or fingerprints) at the surface of the digital skin may prove this assumption wrong. These fine whorly corrugations of the skin strongly modify the short-wavelength stress distribution, as clearly illustrated by the image of a fingertip in contact with a surface (see Fig. 8.1). The regions in between ridges make no contact with the touched surface — to the dismay of the uncareful criminal leaving latent prints on the crime scene — such that the pressure is effectively null in these regions. When the surface is scanned across the fingerprint axis, the periodic modulation of $\bar{\sigma}$ significantly changes the spectral content of the filtering function g, which now displays a large peak at the spatial period of the fingerprints (Fig. 8.7b). This band-pass characteristic of g gives rise to an effective amplification of the subsurface band-pass stress signals induced by textural corrugations of similar wavelength. In the time domain, the selected frequency is $f_0 = v/\lambda$ where v is the finger/substrate relative velocity and λ is the inter-ridge distance. In natural exploratory conditions, v is observed to be of order 10 cm/s and $\lambda \approx 500\,\mu$m, which yields a frequency $f_0 \approx 200$ Hz of the order of the optimal frequency of the Pacinian fibers. Fingerprints thus allow for a conditioning of the texture-induced mechanical signal that facilitates its detection by specific frequency-tuned mechanoreceptors.

3.4 The effect of skin resonance on texture perception

Although the skin can be approximated at small scale as a homogeneous elastic medium, it is better described at large scale as a weakly stretched elastic sheet attached to soft sub-cutaneous tissues. Such a mechanical structure, reminiscent of a drum skin, is expected to display strong resonance properties. Resonant mechanical systems tend to respond with larger amplitude to an oscillatory force signal, when the driving frequency is close to one (or several) particular resonance frequencies. Resonance can be beneficial in sensing systems as it allows for the accumulation of energy over several cycles thus easing the detection of small amplitude signals, provided that they fall within a specific range of frequencies. In a very recent study, Manfredi *et al.* showed that the fundamental resonant frequency of the digital skin is close to the optimal response frequency of Pacinian corpuscles. This observation provides a second possible mechanism that would explain the exquisite sensitivity of these receptors to minute skin vibrations. The authors further showed that local oscillatory stimulations of the fingertip skin at its resonant frequency can propagate, in the form of deflection waves, up to a few centimeters along the finger. They suggested that these traveling waves might drive neural responses in Pacinian corpuscles distant from the contact zone during a tactile exploration, thus enhancing the overall sensitivity of the tactile system to fine textures.

4 Whisker perception in rodents

We now turn to rodents' whisker-mediated tactile perception, by focusing on the rat as this is by far the best studied animal in this domain.

Rats are nocturnal animals that spend a large fraction of their life in narrow tunnels 20–30 cm underground. Their vision is rather limited, both in spectral range and in precision. As a consequence, they rely preferentially on tactile perception, using their long facial hair (called vibrissae or whiskers) to explore their nearby environment or interact with congeners. About 30 whiskers are arranged on each side of the rat's face in regular arrays. When the rat explores an environment where new objects may appear, it frequently agitates its vibrissae back and forth in a rhythmic motion of whisking, which is the analog of the hand movements observed during texture evaluation by humans.

The performance of the whisker system is commonly evaluated through specifically designed behavioral assays. Rats are trained to make a decision regarding the nature or location of one or two objects which they are allowed to palpate in the dark. Such experiments have revealed that rats possess amazing tactile capacities for texture discrimination that may even surpass those of humans. They can, for instance, reliably discriminate a smooth from a rough surface displaying 30 μm grooves spaced at 90 μm intervals, a task that proves difficult for humans. No less surprising is the speed at which they can succeed in such a discrimination task: no more than a few hundred ms of actual whisker/surface contact is generally needed to produce an adequate decision.

The mechanical energy induced upon active contact between the whisker and the object is transduced into neural signals by mechanoreceptors located in the so-called follicle-sinus complex (FSC) in which each whisker is anchored at its base (Fig. 8.8a). Each follicle receives of the order of 200 afferent fibers from the trigeminal ganglion. The majority of these fibers terminate in Merkel cells, similar to the SA1 receptors found in human skin. Other terminations, such as the rapidly adapting lanceolate endings, display electro-physiological responses analogue to the FA1 receptors in humans. The mechanoreceptors of the FSC, that have distant cell bodies in the trigeminal ganglion, send signals to the primary somatosensory cortex (S1), via the trigeminal nuclei located in the brainstem and the thalamus (Fig. 8.8b). In S1, each whisker is represented by a set of neurons called a "barrel". This topographical organisation, where well-delimited regions of the cortex are responsive to a particular whisker stimulation, has made the whisker somatosensory

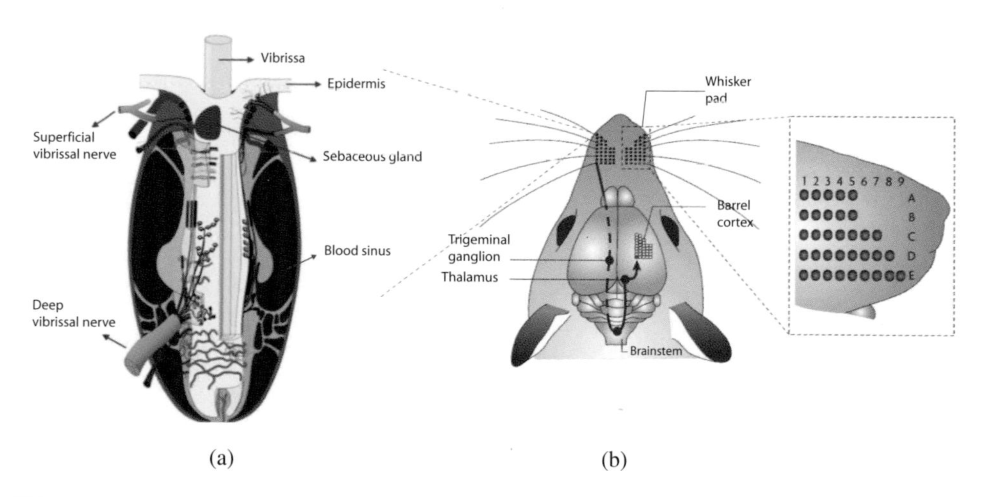

Figure 8.8. The follicle-sinus complex (FSC). (a) Schematic view of the follicle embedded in the blood sinus (dark red). Afferent nerves enter the FSC in its superficial and deep parts. Mechanoreceptors are located at different positions relative to the whisker. Adapted from Diamond and Arabzadeh, *Prog. Neurobiol.* (2013) 103: 28–40. (b) Afferent fibers travel from the FSC to the cell bodies located in the trigeminal ganglion that lies just outside the brainstem. After a synapse in the brainstem, axons of the second-order neurons cross the midline and project to the thalamic somatosensory nuclei. Thalamic neurons in turn project to the barrels in the primary somatosensory cortex. The organization of the vibrissae is somatotopically conserved across the different stages of the somatosensory pathway. Adapted from Diamond *et al.*, *Nat. Rev. Neurosci.* (2008) **9**(8), 601-12 [7].

system a very convenient model for the study of cortical sensory integration. To some extent, this somatotopy is also observed in the brainstem and in the thalamus.

Mechanoreceptors receive inputs from only one whisker and are sensitive to specific stimuli, in a similar way as different classes of cutaneous receptors mediate different types of information. During a typical whisking sequence against an obstacle, a fraction of the FA afferent fibers are found to spike in response to first contact, others to detachment, while a subset of SA mechanoreceptors respond continuously when the whisker is in contact (Fig. 8.9). This specificity likely stems from the intrinsic property of each mechanoreceptor and its localization relative to the whisker within the FSC (Fig. 8.8a). While the FA fibers respond preferentially to stretching and are found in the external sheath of the FSC, SA fibers are more sensitive to compression and are localized closer to the whisker. A fourth category of receptors respond to the whisking movement itself and is thus a potential candidate for encoding the angular position of the whisker with respect to the snout.

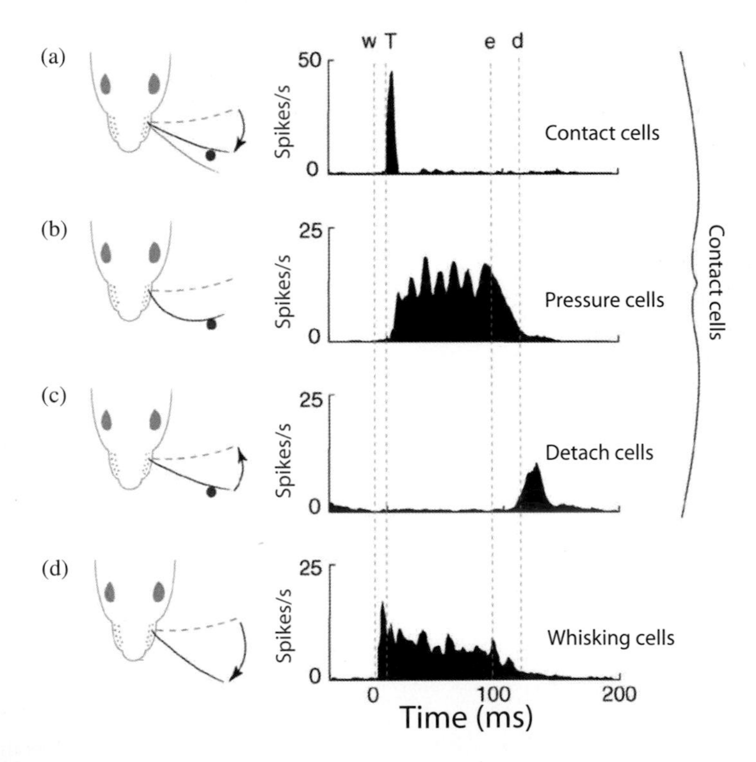

Figure 8.9. Touch cells response in anaesthetized rats. Whiskers are brought into contact with a pole during artificial whisking. Contact (a) and Detach (c) cells show transient response to specific event. (b) Pressure cells tonically respond to contact as long as it is maintained. (d) Whisking cells respond to whisking in itself and are not affected by contact or detach events. Adapted from Knutsen *et al.*, 2009 *Trends in Neurosciences*, **32**, 101–109 [9].

5 The whisker as a transduction line

In the chain of events that lead to the tactile representation of a surface, whiskers play the role of mechanical transducers. They convert displacement and forces imposed along their shaft into stress signals in the FSC that are further transduced into trains of action potentials by mechanoreceptors. As in the case of cutaneous sensing, the characteristics of the FSC deformation that constitutes the relevant mechanical stimulus for the various classes of mechanoreceptors remain mostly unknown. In the absence of such knowledge, it is generally assumed that the bending moment of the whisker within the follicle, and its time derivatives, contain the relevant signal.

Unlike body hair, most mammalian whiskers are conical in shape, very much like fishing lines. There has been a lot of speculation regarding the functional advantage of this tapered shape, that may explain why it has been conserved throughout evolution. One hypothesis relies on the rapid decay of the bending resistance (which scales as the 4th power of the whisker radius) towards the tip. This mechanical property allows the animal to whisk past objects located within a relatively large range of distances without imposing excessive stress on the follicle.

Due to their slender aspect and their elasticity, whiskers exhibit strong mechanical resonance. They tend to oscillate at larger amplitude when driven at specific frequencies in the range of a few tens to a few hundreds of Hertz (Fig. 8.10a). In resonant oscillators, the frequency tuning, i.e. the sharpness of the resonant peak, depends on the degree of damping. For all whiskers, damping ratio is found to be of the order of 0.1, which corresponds to the underdamped regime and thus provides strong frequency selection.

This resonant behavior can be quantitatively understood by modeling the whisker as a tapered elastic rod. Its dynamics can then be obtained by solving the classical Euler–Bernoulli equation, which locally equates the inertial and elastic forces acting on the whisker. The fundamental resonance frequency is found to be proportional to $\alpha c/L$ where L is the length of the whisker, α is the whisker cone angle, and c is the sound wave velocity in the bulk material. Although α and c are essentially conserved across the whisker pad, the length L systematically increases along the anterior-to-posterior axis. The whiskers' resonant frequencies thus span a wide range across the pad (Fig. 8.10b).

5.1 Differential resonance model of texture perception: the tactile cochlea

This spatial map of whisker resonance frequencies forms the basis of a recently proposed tactile encoding scheme for textural information. As the whisker is scanned across a rough surface, its tip is submitted to a fluctuating friction force. Any surface

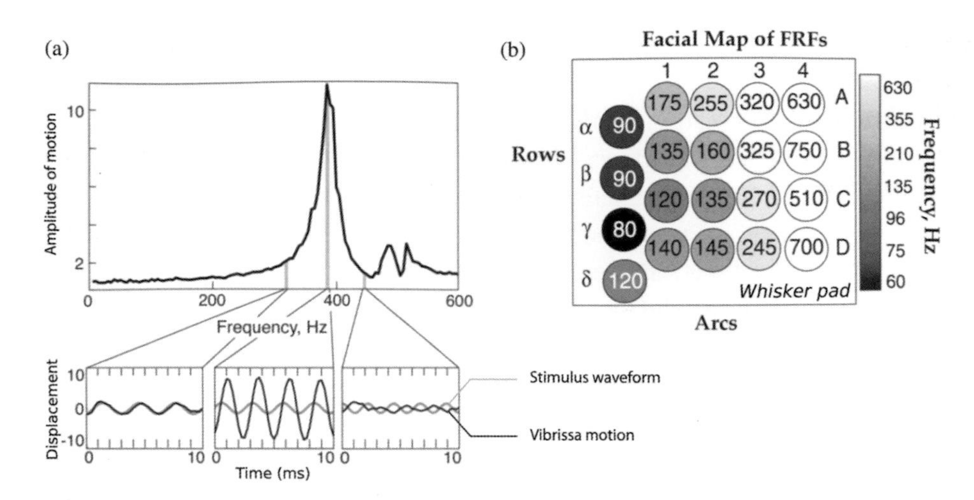

Figure 8.10. Resonance of whiskers. (a) Whisker tip was driven with a sinusoidal waveform of constant amplitude at different frequencies (bottom panels). The amplitude of the whisker motion in response to the stimulation varies. The power spectrum amplitude of the whisker oscillations highlights the resonance frequency of the whisker around 385 Hz (top panel). Adapted from Neimark *et al.*, *J. Neurosci.* (2003) **23**(16), 6499–6509. (b) Map of whisker resonance frequency across the facial whisker array. The antero-posterior gradient of whisker length underlies a somatotopic map of resonance frequency, with higher frequencies observed for short anterior whiskers. Adapted from Andermann *et al.*, *Neuron* (2004) **42**, 451–463.

corrugation of wavelength l drives whisker oscillations at a frequency v/l where v is the scanning speed. A particular spectral component of the surface topography will thus elicit a significant whisker oscillation only when the corresponding frequency falls within the whisker's resonant range. The complex topographical information is decomposed in a series of modes, each of them being transduced by specific whiskers. In this model, the texture identity is spatially encoded in the particular pattern of whiskers activated during tactile exploration. This tonotopic processing scheme, analog to that at play in auditory perception, would make the whisker pad the equivalent of a tactile cochlea (see Chapter 7).

5.2 The kinetic signature model

This model is currently disputed. Although the whisker vibrations, measured during actual tactile exploration phases, clearly contain a dominant resonant mode, its amplitude seems weakly dependent on the nature of the palpated texture. The reason is that in most realistic configurations, the motion of the whisker tip along the surface is not smooth. Rather than building-up continuously, resonant oscillations occur during brief episodes elicited by stick-slip events at the frictional contact between the whisker and the probed surface. The whisker tip gets pinned on surface defects, then suddenly slides, releasing a burst of elastic energy that triggers a series of large oscillations (ringing).

A competing model of texture encoding has thus been proposed where texture discrimination would rely on the timing, number and/or energy of these discrete events. Each stick-slip process would trigger a particular sequence of mechanical stimulation in the FSC, a *kinetic signature* to which a population of cortical neurons would be specifically tuned. In recent experiments, where whisker trajectories were monitored while simultaneously recording neurons in the S1 regions of the cortex, a significant fraction of neurons were found to actually respond specifically to these high-acceleration events (Fig. 8.11). The peak acceleration and number of these discrete events were found to consistently vary with the surface roughness when sandpapers were used as a stimulus. Unlike the differential resonance model, this hypothesis does not require an extended time of integration and is thus consistent with the observation that rats can discriminate textures through very brief exploration sequences.

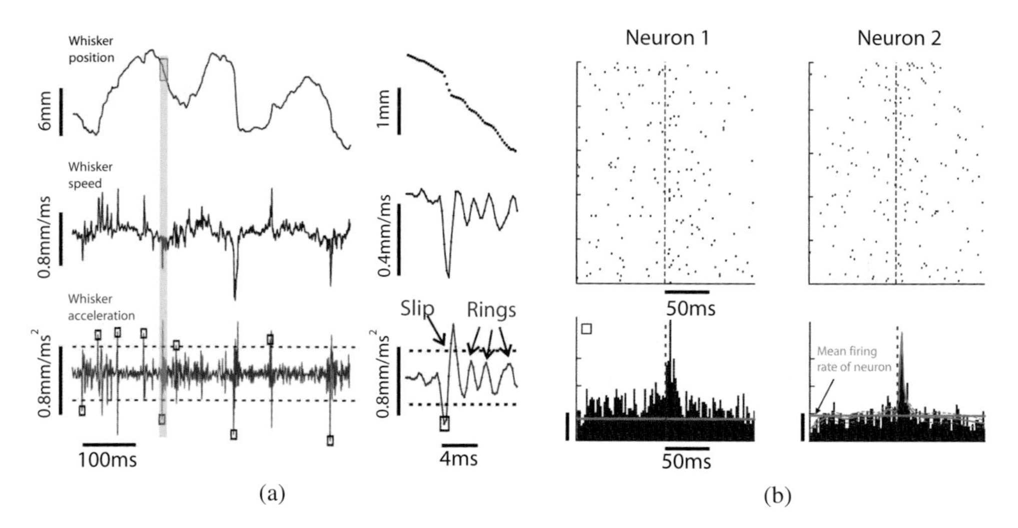

Figure 8.11. Stick-slip dynamics and cortical response. (a) Whisker motion during whisking on sandpaper (boxes are slips detected by acceleration threshold crossings; ringing oscillations are discarded). Gray time period is expanded in the right panels. (b) Neuronal response of cortical neurons aligned on slip events. Top: Spike rasters for two slip-responsive neurons aligned to slip events. Bottom: Post-stimulus time histogram aligned to slips only (left) and slips and ringings (right) for the 2 neurons shown above. Slips drive precisely timed spikes in S1 neurons. Orange line denotes the mean firing rate of the neuron averaged over the entire recording session. Adapted from Jadhav *et al.*, *Nat. Neurosci.* (2009) **12**(6): 792–800.

6 Conclusion

Whisker- and fingertip-mediated perception of texture share important similarities. They both rely, for best efficiency, on voluntary movements through stereotypical exploratory procedures. These active processes convert the spatial topography of the texture into a particular kinetic of the tactile organ that is encoded by peripheral mechanoreceptors. Different non-neural mechanisms may play a role in shaping the relevant stimulus, enhancing the sensitivity and discrimination capabilities of the sensory system. In the fingertip, epidermal ridges confer to the skin the property of a band-pass mechanical transducer, with an optimal frequency that can be matched by the individual to that of the PC receptors. In both systems, the organ mechanical resonance permits the focus of mechanical energy within a narrow bandwidth, thus easing the detection by frequency-tuned receptors. Strong differences however exist between these two sensory systems. The spatial encoding of coarse texture by populations of SA1 receptors does not seem to have an equivalent in the whisker system. Similarly, stick-slip processes, which seem to play a central role in whisker-mediated texture perception, are generally avoided in human tactile evaluation of texture.

Thanks to these various biomechanical tricks, the human hand and the rodent's whisker pad are extraordinary sensitive tools with no equivalent in man-made equipments. This might change in the future with the rapid development of biomimetic systems that aim at implementing similar strategies in the tactile appendages of humanoïd or rodent-like robots.

Bibliography

[1] Mathew E. Diamond. Texture sensation through the fingertips and the whiskers. *Curr. Opin. Neurobiol.* (2010) **20**(3), 319–327.

[2] Mark Hollins and Sliman J. Bensmaia. The coding of roughness. *Can. J. Exp. Psychol.* (2007) **61**(3), 184–195.

[3] Shantanu P. Jadhav and Daniel E. Feldman. Texture coding in the whisker system. *Curr. Opin. Neurobiol.* (2010) **20**(3), 313–318.

[4] Lynette A. Jones and Suzan J. Lederman. *Human Hand Function*. Oxford University Press (2006).

[5] Louise R. Manfredi, Andrew T. Baker, Damian O. Elias, John F Dammann III Mark C. Zielinski, Vicky S. Polashock, and Sliman J. Bensmaia. The effect of surface wave propagation on neural responses to vibration in primate glabrous skin. *PLoS One* (2012) **7**(2), e31203.

[6] Jason Wolfe, Dan N. Hill, Sohrab Pahlavan, Patrick J. Drew, David Kleinfeld, and Daniel E. Feldman. Texture coding in the rat whisker system: slip-stick versus differential resonance. *PLoS Biol.* (2008) **6**(8), e215.

[7] Mathew E. Diamond, Moritz von Heimendahl, Per Magne Knutsen, David Kleinfeld, and Ehud Ahissar. 'Where' and 'what' in the whisker sensorimotor system. *Nat. Rev. Neurosci.* (2008) **9**(8), 601–612.

[8] Yves Boubenec, Daniel E. Shulz, and Georges Debrégeas. An amplitude modulation/demodulation scheme for whisker-based texture perception. *J. Neurosci.* (2014) **34**(33), 10832–10843.

[9] Per Magne Knutsen, David Kleinfeld, and Ehud Ahissar. Orthogonal coding of object location. *Trends Neurosci.* (2009) **32**(2), 101–109.

[10] Julien Scheibert, Sébastien Leurent, Alexis Prevost, and Georges Debrégeas. The role of fingerprints in the coding of tactile information probed with a biomimetic sensor. *Science* (2009) **323**(5920), 1503–1506.

[11] Shantanu P. Jadhav, Jason Wolfe, and Daniel E. Feldman. Sparse temporal coding of elementary tactile features during active whisker sensation. *Nat. Neurosci.* (2009) **12**(6), 792–800.

9

Intermittent search strategies

Olivier Bénichou, *CNRS researcher, Laboratoire de Physique Théorique de la Matière Condensée, Université Pierre et Marie Curie.*
Raphael Voituriez, *CNRS researcher, Laboratoire Jean Perrin and Laboratoire de Physique Théorique de la Matière Condensée, Université Pierre et Marie Curie.*

1 Introduction

How long does it take a "searcher" to find a "target" of unknown location? This apparently simple question arises in various contexts, such as the search for victims in avalanches, the search for food by animals, or at the microscopic scale the search for a target sequence on DNA by a regulatory protein. In all these examples, the search time is a limiting quantity that often needs to be optimized.

The question of search optimizing strategies was formalized early by US-Navy mathematicians during World War II: the purpose was then to locate as fast as possible the enemy submarines. The problem was to determine the optimal spatial distribution of searching ships, compatible with a given likelihood map of the target position. More recently, these methods have been generalized in the context of rescue operations by coast guards, and also employed for example to recover the wreck of the submarine USS Scorpio in 1968. However, such methods are useful only when at least a partial information about the target location is available.

In the opposite case where no information on the target location is available so that no specific region should be favored, the searcher can adopt a random trajectory. Such random trajectories are also involved usually at the microscopic scale when the searcher is subject to thermal fluctuations, as in the case of a reactive molecule searching for its reaction partner. These so-called "random search strategies" have been the subject of an increasing number of publications in statistical physics over the recent years, including influential works from the group of G. Viswanathan [1].

These authors consider a searcher that is able to detect targets *while it moves*, and whose trajectory is made of a sequence of straight ballistic moves of random direction and random length l. Viswanathan and coworkers studied the case where the probability distribution $p(l)$ is a Lévy law defined by $p(l) \propto l^{-\mu}$, where $\mu > 1$ is a given parameter. In the particular case of so-called "revisitable" targets, which immediately reappear at the same position after each visit of the searcher, they showed that the Lévy law that minimizes the search time corresponds to $\mu \approx 2$.

This situation of revisitable targets is however rather artificial since in most of real situations targets disappear once they are discovered: this is for instance the case of a prey captured by a predator. In this last more realistic situation, which we will consider in this chapter, the model of Viswanathan and coworkers leads to an expected result: the search time is minimized by a simple uninterrupted ballistic motion. We present here search strategies that are alternative to the celebrated Lévy strategies, which in fact are optimal only in the particular case of revisitable targets.

More precisely, we address the following question: what happens if the searcher is not able to detect targets *while it moves*? This situation is quite common since in general motion degrades perception abilities. Often, motion and detection are in fact intrinsically incompatible: this is the case with targets hidden in the ground, as would be the case of keys lost in a sandy beach (Fig. 9.1). Recent works put forward a new type of random search, exemplified by the searcher for keys (Fig. 9.1), called intermittent search strategies: the searcher combines phases of careful inspection to detect the target and phases of rapid motion during which target detection is not possible. As we shall see, intermittent search strategies are widely observed in

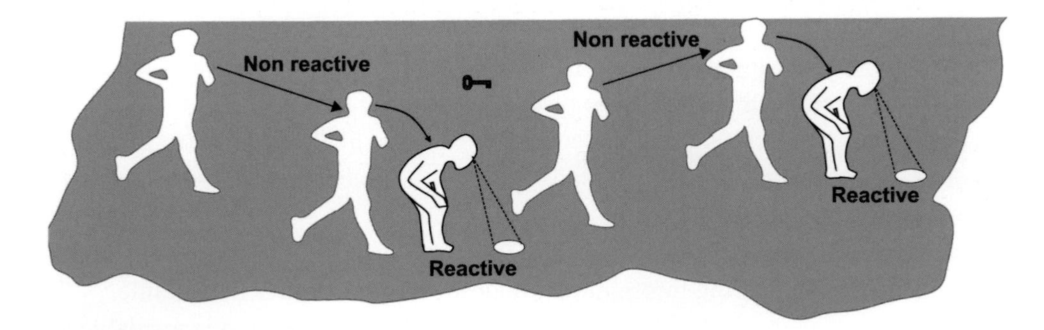

Figure 9.1. Intermittent search strategy: the searcher alternates phases of careful inspection to detect the key, and phases of fast motion during which he is unable to detect the target.

nature at various scales. One can then suggest that the reason why such strategies are so common in nature is because they are *extremely efficient*.

2 Animals looking for food

The target search problem is especially relevant in the context of animals looking for food. The efficiency of this search behavior is of vital importance for the animal, and one can suggest that fast behaviors have been selected by evolution. Analyzing the search behavior of animals and assessing its efficiency is therefore very important, and behavioral ecologists have embarked in the study of the behaviors of animals looking for food. Such studies schematically put forward two extreme types of search strategies. The first one corresponds to predators that search for preys while they move. It is the case for example of pelagic fish like tuna that never stop swimming. The second strategy, called ambush strategy, corresponds to animals such as rattle snakes, which stay immobile for long periods of time, waiting for a prey. These two strategies in fact should be considered as extreme situations.

A few decades ago, O'Brien and colleagues showed that in fact the search behavior of many species couldn't be described by these extreme strategies. They observed, first on the example of planktivorous fish and next on examples of lizards and bird species, that many animals in fact displayed an intermediate behavior, named "saltatory", defined by the following observations: the animal alternates phases of fast motion and "stationary" phases, during which the animal moves much less, but during which targets can be detected. The duration of each of the phases can vary significantly from species to species, but observations reveal the existence of correlations between the times spent in each phase, independent of the size or weight of the animal.

To explain such observations, behavioral ecologists have developed models based in general on the idea of maximization of the energy gain: the energy provided by food consumption should compensate at best the energy lost during the search for preys. Most of these models involve parameters that are difficult to access experimentally. In addition, they are specific to a given species. Is it possible to propose a different approach relying on a generic modeling of such saltatory search behaviors?

3 Model of intermittent search

A model of random search accounting for available experimental observations has been proposed recently [2]. This model relies directly on the idea of intermittent

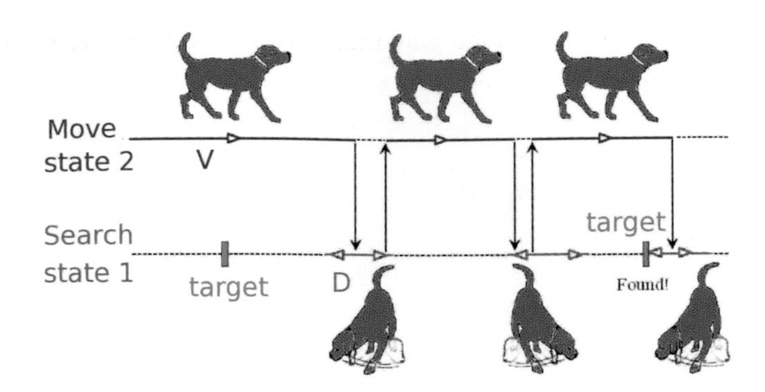

Figure 9.2. The two-state model of intermittent search.

search suggested in the example of the key searcher. The searcher alternates between two states (Fig. 9.2): in state 1, the searcher hardly moves and carefully scans the environment, while in state 2 it is unable to detect targets but moves fast. A key quantity in the problem of animals searching for food is the *search time* for a target. The search strategy that minimizes the search time can then be called optimal, since it enables the searcher to get food rapidly and outperform its competitors.

Let us consider more carefully the rules defining the model. First, available experimental observations indicate that after each stationary phase the animal in practice only slightly modifies its direction of motion, which suggests a 1-dimensional model. Second, state 2 is described as a ballistic phase of constant velocity v. This in fact corresponds to the most efficient way of exploring space. We also make the assumption that the time spent in this state 2 is a random variable distributed according to an exponential law of mean τ_2. The time spent in state 1 is also taken as exponentially distributed, of mean τ_1, and we hypothesize in addition that this state 1 can be described as a Brownian motion of diffusion coefficient D. Such hypothesis calls for comments. First, the model obviously does not aim at describing precisely the detection phase, which involves very complex biological mechanisms, but only at extracting the main characteristics of these mechanisms in order to estimate the time needed to detect a target. Second, we stress that it is not necessarily the very position of the searcher that performs diffusion, but rather the focus point of the detection system, such as the focus point of the eyes. Last, we underline that this process of visual search is often described in terms of random walks, which makes our hypothesis in agreement with previous studies.

4 Minimizing the search time

Can one define an optimal strategy? In other words, is it possible to adjust the times τ_1 and τ_2 spent in each state to minimize the search time?

The search time can actually be calculated. A striking feature of the result is that, in the limit of low target density, the search time is proportional to the mean distance L between targets, and not L^2 as would be the case if the diffusive phases were not interrupted by ballistic phases. This reflects the efficiency of intermittent strategies as compared to purely diffusive strategies.

Concerning the minimization of the search time, a detailed analysis leads to distinguishing two cases. If the animal spends more time searching than moving, the model predicts that the times spent in each state are related by a power law: the time τ_2 that minimizes the search time is proportional to $\tau_1^{2/3}$. One can also consider the case of animals that spend more time moving than searching: in that case one still obtains a power law, but now the search time is minimized when τ_2 is proportional to $\tau_1^{3/5}$.

Available experimental data provide for a set of 18 animal species, including lizards, fish and birds the mean times spent in each of the two phases. The comparison of these experimental data with the power laws predicted by the theoretical analysis yields a satisfactory agreement (cf. Fig. 9.3). Hence, it appears that animals

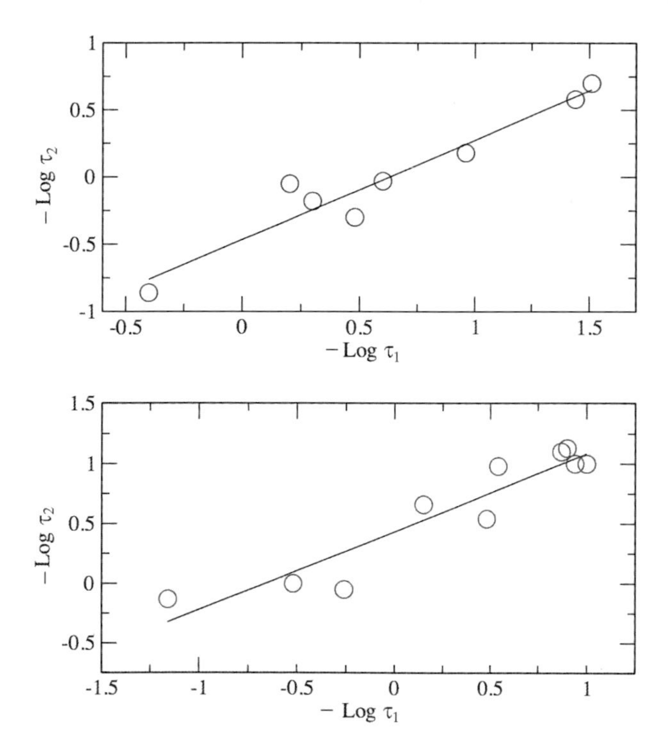

Figure 9.3. Plots (in log/log scale) of the mean time τ_2 spent in the fast moving phase as function of the mean time τ_1 spent in the careful scanning phase. The top panel corresponds to animals that spend more time moving than searching, and the bottom panel to animals that spend more time searching than moving. Each point stands for given species; lines are linear fits and yield the slopes 0.7 (top) and 0.6 (bottom).

as various as fish, lizards and birds perform intermittent search strategies to minimize their search time for preys.

5 Should animals really perform Lévy strategies?

The question of optimizing search strategies in the context of behavioral ecology has recently been very controversial. The status of Lévy walks as optimal search strategies has in fact been questioned in 2006 from a theoretical point of view [3], essentially because in most of realistic cases the optimal strategy predicted by the Lévy walk model is trivially the straight ballistic motion. Since then the Lévy walk model has also been questioned experimentally. In fact, it was shown [4] that most of the data that had been used to identify Lévy walks in animal trajectories were misinterpreted, and most likely were consistent with classical exponential random walks. Finally, as opposed to what is commonly stated, animals have no reason to perform Lévy strategies, since they are not optimal. In turn, intermittent strategies, which can really be minimized, constitute an alternative model of animal trajectories.

6 How does a protein find its target sequence on DNA?

Target search problems are also involved at much smaller scales, such as the scale of a single molecule. In fact, a chemical reaction can be seen as the search of a reactive site by a reactant. Living cells provide striking examples of chemical reactions whose kinetics must be optimized in order to efficiently fulfill a biological function. A prototypical example is given by proteins looking for specific target sequences on DNA. It is the case of transcription factors, which bind to specific DNA sequences in order to regulate — either promote or repress — the transcription of a given gene. Another well known example is provided by the restriction enzyme EcoRV, which enables bacteria to fight viral attacks: EcoRV is designed to cut the viral DNA at a specific site as soon as it is introduced in the bacterium, thus preventing its replication. Such reactions involve very short target sequences — typically a few base pairs long. Given that such targets are sometimes unique, from the point of view of the protein the search problem is therefore very hard, and a simple diffusive strategy, as usually observed in chemistry, would yield search times of the order of hours, far longer than what is observed *in vivo* (of the order of seconds). This paradox has long remained unresolved.

Recently, the development of new experimental techniques such as the micro manipulation of single DNA molecules or the tracking of fluorescent proteins at the single molecule level have shown that two mechanisms can be involved in the process of target search by proteins: a slow diffusive motion along the DNA molecule during which the protein remains non specifically bound, and a much faster free diffusive motion in the surrounding medium. The protein can then switch randomly between these two phases by binding and unbinding DNA, as was early suggested by O. Berg and collaborators [5]. In fact, proteins such as EcoRV have a site that is locally positively charged, which induces an affinity for DNA, which is negatively charged. The binding of such proteins to DNA is therefore energetically favored, and is called non-specific since it is largely independent of the sequence. Thermal fluctuations, that are due to random collisions with the molecules of the surrounding medium, are then responsible for the diffusive motion of the protein along the DNA, and sometimes for its unbinding. This type of trajectories is in fact intrinsically intermittent, and can be compared to our first example of key searcher. The protein carefully scans the DNA while it is bound, and moves much faster but without any chance to detect the target when it is not bound to DNA.

A theoretical model similar to the one introduced in the context of animals looking for food, except for the fact that fast phases are described by a random walk (Fig. 9.4) to model the diffusive excursions of the protein in the surrounding medium, shows that the search time can be minimized if the protein spends

Figure 9.4. Artistic view of a protein looking for a target site on DNA. Artwork by Virginie Denis, Pour la Science n° 352, February 2007.

the same amount of time bound to the DNA and free in the solvent. The efficiency of an intermittent strategy can be discussed in this example: in the limit where the protein never unbinds from DNA, the search time scales as L^2 (L being the length of the DNA molecule), which becomes extremely large in the relevant limit of long DNA molecules. However, the model shows that the intermittent search strategy yields a search time that scales as L, which is hence much smaller in the large L regime than in the case of a purely diffusive strategy. By comparing the theoretical model to experimental data, it can be shown that the typical DNA length that is scanned by the protein during a single binding phase is of the order of 300 base pairs, and therefore much shorter than the total length of a DNA molecule (which is typically larger than 10^6 base pairs). This very short length of diffusive excursions along the DNA has since then been measured directly by analyzing trajectories of single EcoRV molecules on DNA by the team of P. Desbiolles at ENS Paris [7, 8], which confirms that the restriction enzyme EcoRV makes use of the extreme efficiency of intermittent strategies to find its target on DNA.

7 Active intermittent transport in cells

At the microscopic scale, intermittent search strategies are also observed in the context of active transport of vesicles (or other molecular assemblies) in cells. Because of the intrinsically out of equilibrium nature of the cellular medium, the motion of a typical tracer particle in a cell can be very different from the usual thermal diffusion. In fact, in a cell, a tracer particle such as a vesicle can diffuse, but it can also bind to molecular motors, which thanks to ATP hydrolysis (the chemical energy of the cell) are able to perform a ballistic directional motion along the cytoskeleton filaments (Fig. 9.5). Such motion, called active, is indeed intermittent in the general sense defined above, since it involves a slow phase, here thermal diffusion, and a fast phase of ballistic motion induced by molecular motors. We assume here that the tracer is not reactive when it is bound to motors, as in the examples where the targets are membrane proteins. Such intermittent trajectories have been observed at the single vesicle level at IBPC in Paris [9], and these experiments indeed indicate that reaction (in that case exocytosis of the vesicle) occurs mostly in the diffusive phase. In a general context, we aim here at discussing the impact of active transport on reaction kinetics in cells. For that purpose, we estimate the rate constant of a simple first order chemical reaction, which is given by the inverse mean first-passage time to the target.

8 Optimizing the kinetic rate constant

In the case of a tracer moving in 3 dimensions, for example in the volume V of the cytoplasm (Fig. 9.5), we have shown that intermittent transport is more efficient than a simple diffusion of diffusion coefficient D, and that it makes possible the

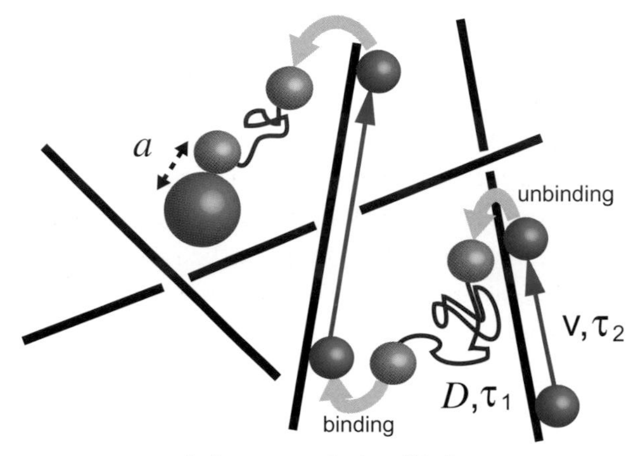

Active transport in the cell bulk.

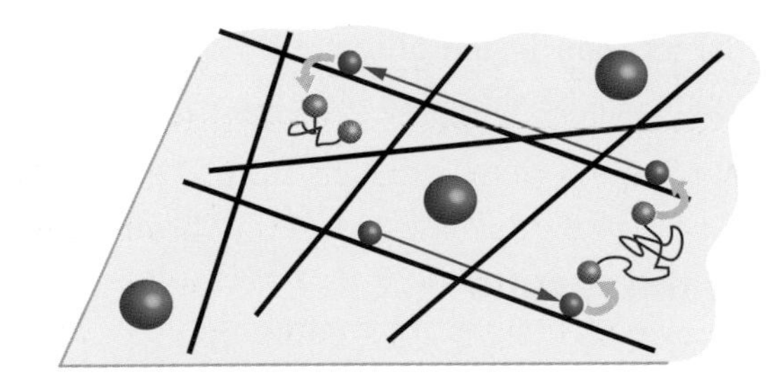

Active transport on 2-dimensional structures such as membranes.

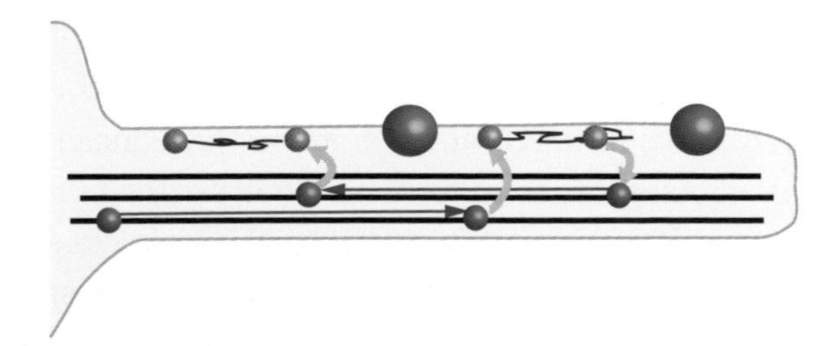

Active transport along 1-dimensional structures such dendrites or axons.

Figure 9.5. Active transport in cells: different relevant geometries.

maximization of the kinetic rate constant as soon as the reaction radius (distance below which reaction can take place) verifies $a > a_c \simeq 6.5\, D/V$. In usual cellular conditions, D is comprised between $10^{-2}\ \mu\text{m}^2\,\text{s}^{-1}$ for vesicles and $10\ \mu\text{m}^2\,\text{s}^{-1}$ for small proteins, and the typical velocity of a molecular motor is of the order of $v \sim 1\ \mu\text{m}\,\text{s}^{-1}$. The critical reaction radius is then comprised between 10 nm, which is smaller than any cell organelle, and 10 μm, which is comparable to the cell size. For tracers of the size of a vesicle, active transport therefore enables an optimal exploration of space, whereas for individual molecules simple diffusion is more efficient. In addition, it should be noted that in usual cellular conditions the optimal strategy is realized in the case of a reaction radius of 0.1 μm when the typical interaction time with molecular motors is of the order of 0.1 s. This value is compatible with experimental observations and suggest that intracellular transport could be close to optimal efficiency. Lastly, it can be shown that in these conditions the optimal kinetic rate constant can be up to 10 times larger than in the absence of active transport.

In the case of a tracer evolving in a 2-dimensional space, typically along a membrane, or in a one dimensional structure such as axons or dendrites (Fig. 9.5), the kinetic rate constant can also be maximized, and in fact the optimal rate constant can become much larger. In one dimension, for a low enough concentration of targets, it can be 100 times larger than in the absence of active transport, since in low spatial dimension the gain can be shown to increase when the target concentration is decreased.

9 Robustness of the results

The previous examples show that intermittent search strategies are observed at various scales. This suggests that intermittent search is a general mechanism, which enables the optimization of the success of random searches for hidden targets. To support this hypothesis, a minimal model has been proposed, which only relies on the essential ingredients of intermittent strategies introduced up to now on the example of animals looking for food, proteins looking for a target sequence on DNA, or of active intermittent transport in cells, and which avoids the specificities of these examples.

Following the above examples, we consider a 2-state searcher, which alternates between phases of slow but reactive motion (which enable target detection) (state 1) and phases of fast but non-reactive motion (state 2). The random durations of each phase i are assumed to be exponentially distributed of mean τ_i, which means that the searcher has no memory in time: it does not remember when the last transition between phases occured. The fast phase (state 2) is described as a ballistic motion

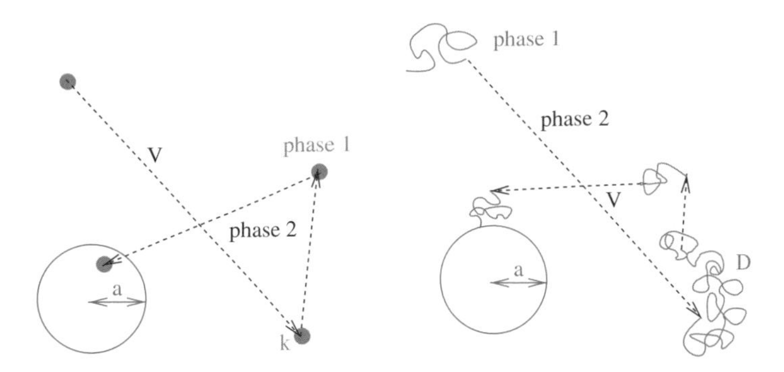

Figure 9.6. A minimal search model: the searcher alternates between detection phases and motion phases during which targets cannot be detected. Left: static mode of detection. Right: dynamic mode of detection.

of constant velocity v, performed in a direction drawn at random. Hence, one can consider that the searcher has no memory in space either. In that sense it can be considered as minimal, since it has minimal cognitive abilities.

In order to assess the robustness of this minimal model, we consider two limiting modes of target detection (Fig. 9.6). The first one, called "dynamic" mode, is a direct generalization of the model introduced to describe animals looking for food: the detection phase is assumed to be *diffusive*, characterized by a diffusion coefficient D. In addition reactivity is assumed to be infinitely efficient, in the following sense: as soon as the target is reached in state 1, it is found. The second one, called "static" mode, is very different. During detection phases, the searcher is assumed *immobile*, but reactivity is now imperfect: the target is detected only with finite probability if it is within the reach of the searcher.

It can be shown [3] that for each of these two detection modes, the search time can be minimized as a function of the mean durations of each of the phases 1 and 2. This result holds for any spatial dimension $d = 1, 2, 3$. In other words, there exists a unique optimal way of splitting the time spent in phases 1 and 2 to find the target as fast as possible. Remarkably, even if the searcher is assumed to have minimal cognitive abilities, the intermittent strategy allows the searcher to very significantly reduce the search time, as compared to the case where the searcher would always stay in phase 1. Typically, one can show in the dynamic case that the search time can be reduced by a factor av/D, which can become very large if the velocity v is sufficient. A striking feature is that the optimal strategy is realized for both detection modes, for the *same* value of the mean time spent in the moving phase 2. This quantity is therefore independent of the specificities of the detection mechanism and can therefore be considered as a "universal " feature of intermittent search processes, which only depends on the space dimension.

10 Conclusion

Finally, we have shown that intermittent search strategies are efficient search strategies that enable the minimization of the search time for hidden targets. Hence, even if it may seem counterintuitive, the key searcher should indeed interrupt his phases of careful search to perform fast displacement phases. Even if they may seem to be taking up more time, such "blind" phases of displacement can in fact speed up the search process. In addition we have seen that a model searcher with minimal cognitive abilities can significantly improve its search efficiency by tuning the characteristic times it spends in each of the careful search and displacement phases. Such strategy seems easily implementable since it does not involve potentially costly cognitive skills. It therefore seems favorable from an evolutionary point of view, which could explain the observation of such intermittent strategies in situations as various as animals looking for food or proteins looking for a target sequence on DNA.

Intermittent search strategies could also be used in the context of human search tasks, such as the search for explosives or the search for victims in avalanches or earthquakes, as suggested by M. Schlesinger from the research department of the US navy [11].

Bibliography

[1] Viswanathan GM, Buldyrev SV, Havlin S, da Luz MGE, Raposo EP, Stanley HE. *Nature* (1999) **401**, 911.

[2] Bénichou O, Coppey M, Moreau M, Suet P-H, Voituriez R. *Phys. Rev. Lett.* (2005) **94**, 198101.

[3] Bénichou O, Loverdo C, Moreau M, Voituriez R. *Phys. Rev. E* (2006) **74**, 020102(R).

[4] Edwards AM, Phillips RA, Watkins NW, Freeman MP, Murphy EJ, Afanasyev V, Buldyrev SV, da Luz MGE, Raposo EP, Stanley HE, Viswanathanand GM. *Nature* (2007) **449**, 1044.

[5] Berg OG, Winter RB, von Hippel PH. *Biochemistry* (1981) **20**, 6929.

[6] Coppey M, Bénichou O, Voituriez R, Moreau M. *Biophys. J.* (2004) **87**, 1640.

[7] Bonnet I, Biebricher A, Porté P-L, Loverdo C, Bénichou O, Voituriez R, Escudé C, Wende W, Pingoud A, Desbiolles P. *Nucleic Acid Res.* (2008) **36**, 4118.

[8] Loverdo C, Bénichou O, Voituriez R, Biebricher A, Bonnet I, Desbiolles P. *Phys. Rev. Lett.* (2009) **102**, 188101.

[9] Huet S, Karatekin E, Tran VS, Fanget I, Cribier S, Henry J-P. *Biophys. J.* (2006) **91**, 3542.

[10] Loverdo C, Benichou O, Moreau M, Voituriez R. *Nat. Phys.* (2008) **4**, 134.

[11] Schlesinger MF. *Nature* (2006) **443**, 281.

Index